Fire and Polymers

ACS SYMPOSIUM SERIES **797**

Fire and Polymers

Materials and Solutions for Hazard Prevention

Gordon L. Nelson, Editor
Florida Institute of Technology

Charles A. Wilkie, Editor
Marquette University

American Chemical Society, Washington, DC

Library of Congress Cataloging-in-Publication Data

Fire and polymers : materials and solutions for hazard prevention / Gordon L. Nelson, editor, Charles A. Wilkie, editor.

    p. cm.—(ACS symposium series ; 797)

    "Developed from a symposium sponsored by the Division of Polymeric Materials: Science and Engineering at the 220th National Meeting of the American Chemical Society, Washington, D.C., August 20–24, 2000."

    Includes bibliographical references and index.

    ISBN 0–8412–3764–6

    1. Polymers—Fire testing—Congresses. 2. Fire resistant polymers—Congresses.

    I. Nelson, Gordon L., 1943- II. Wilkie, C. A. (Charles A.) III. American Chemical Society, Division of Polymeric Materials: Science and Engineering. IV. American Chemical Society. Meeting (220th : 2000 : Washington, D.C.) V. Series.

TH9446.5.P65 F553   2001
628.9′223—dc21   2001022866

The paper used in this publication meets the minimum requirements of American National Standard for Information Sciences—Permanence of Paper for Printed Library Materials, ANSI Z39.48–1984.

PRINTED IN THE UNITED STATES OF AMERICA

Foreword

The ACS Symposium Series was first published in 1974 to provide a mechanism for publishing symposia quickly in book form. The purpose of the series is to publish timely, comprehensive books developed from ACS sponsored symposia based on current scientific research. Occasionally, books are developed from symposia sponsored by other organizations when the topic is of keen interest to the chemistry audience.

Before agreeing to publish a book, the proposed table of contents is reviewed for appropriate and comprehensive coverage and for interest to the audience. Some papers may be excluded to better focus the book; others may be added to provide comprehensiveness. When appropriate, overview or introductory chapters are added. Drafts of chapters are peer-reviewed prior to final acceptance or rejection, and manuscripts are prepared in camera-ready format.

As a rule, only original research papers and original review papers are included in the volumes. Verbatim reproductions of previously published papers are not accepted.

ACS Books Department

Contents

Non-Halogen Fire Retardants

Halogen

Assessment and Performance

Indexes

Preface

In the United States we each have about a 40% probability lifetime chance of being involved in a large enough fire to cause the fire department to arrive at our door. 1.8 million fires occur in the United States annually, resulting in $10 billion in damage. Although the United States has one of the highest rates of fire in the world, fire is a worldwide problem. Most fires involve the combustion of polymeric materials. Flame retardants are the largest single class of additives used in plastics. Flame retardants alone constitute a $1 billion business worldwide. Thus, a clear need exists for a peer-reviewed book on the latest topics in fire science from a materials perspective.

Because fire and polymers are an important social issue and because of the interest in and the complexity of fire science, a symposium and workshop were organized in conjunction with the 220th American Chemical Society (ACS) National Meeting in Washington, D.C. The symposium builds upon previous symposia in 1989 and 1994. Forty-eight papers from leading experts in ten countries were presented. From those, 28 papers were carefully selected for incorporation into this volume.

Acknowledgments

We gratefully acknowledge the ACS Division of Polymeric Materials: Science and Engineering and the Petroleum Research Fund for support of travel for international academic speakers at the symposium. We also acknowledge Cyndi Johnsrud for her extensive assistance with the myriad of details in preparation for the symposium and workshop and in the preparation and organization of PMSE preprints from the symposium.

Gordon L. Nelson
College of Science and Liberal Arts
Florida Institute of Technology
150 West University Boulevard
Melbourne, FL 32901—6975

Charles A. Wilkie
Department of Chemistry
Marquette University
P.O. Box 1881
Milwaukee, WI 53201—1881

xi

Chapter 1

Fire Retardancy in 2001

Gordon L. Nelson[1] and Charles A. Wilkie[2]

[1]Florida Institute of Technology, 150 West University Boulevard,
Melbourne, FL 32901-6975
[2]Department of Chemistry, Marquette University, P.O. Box 1881,
Milwaukee, WI 53201-1881

Fire is a world-wide problem which claims lives and causes significant loss of property. Some of the problems are discussed and the solution delineated. This peer-reviewed volume is designed to be as the state-of-the-art. This chapter provides a perspective for current work.

In the United States every 17 seconds a fire department responds to a fire somewhere in the nation. A fire occurs in a structure at the rate of one every 60 seconds. A residential fire occurs every 82 seconds. Fires occur in vehicles every 85 seconds. There is a fire in an outside property every 34 seconds. The result is 1.8 million fires per year attended by public fire departments. In 1999 those fires led to $10 billion in property damage and 3,570 civilian fire deaths (one every 147 minutes) and 21,875 injuries (one every 24 minutes). Some 112 fire fighters died in the line of duty. Fires have declined over the period 1977 to 1999, most notably structural fires, from 1,098,000 to 523,000. Civilian fire deaths in the home (81% of all fire deaths) declined from 6,015 in 1978 to 2,895 in 1999. While those declines are progress, the United States still maintains one of the highest rates of fire in the world. One of the remaining needs is to make home products more fire safe (1-4).

The higher rate of fire in the United Sates versus most industrial nations is

2

likely a product of five factors: (1) The U.S. commits fewer resources to fire prevention activities; (2) there is greater tolerance in the U.S. for "accidental" fires; (3) Americans practice riskier and more careless behavior than people in other countries (use of space heaters for example); (4) homes in the U.S. are not built with the same fire resistance and compartmentation as in some other countries; and (5) perhaps most importantly, people in the U.S. have more "stuff" than those in other countries (i.e., higher fire load) as well as a higher number of ignition sources (higher use of energy).

Polymers form a major part of the built environment, of the materials around us. Fire safety thus, in part, depends upon those materials. Polymers are "enabling technology," thus advances in numerous technologies depend upon appropriate advances in polymers for success. While polymers are both natural and synthetic, this book largely focuses on the fire aspects of synthetic polymers. Consumption of the five major thermoplastic resins—low-density polyethylene, high-density polyethylene, polyvinyl chloride, polypropylene, and polystyrene—is over 100 million metric tons worldwide. These resins constitute 80% of thermoplastics sold and about 55% of all synthetic polymers sold, including thermoplastics, thermosets, synthetic fibers, and synthetic rubbers. The U.S. constitutes about 25% of worldwide plastics consumption, the European Union about 22%, and Japan about 9% (5). In Table I are presented plastics production statistics in the U.S. for 1999 (6). Table II provides major market application statistics (7). Figure 1 shows the growth in the use of plastics from 1994 through 1999 (8).

All organic polymers are combustible. They decompose when exposed to heat, decomposition products burn, smoke is generated, and the products of combustion are highly toxic, even if only CO and CO_2. Fire performance is not a single material property (if it is a material property at all). Fire performance combines thermal decomposition, ignition, flame spread, heat release, ease of extinction, smoke obscuration, toxicity, and other properties. A regulation utilizes those tests and assessments for materials and systems used in and appropriate to a particular application. Thus, for example, it is appropriate for small appliances to only worry about ignitability by a Bunsen burner flame or a needle flame, since from an internal point of view that is the size of source possible in real life appliance failures.

This volume is about the latest research in the field of fire and polymers. Much work continues focused on improving the fire performance of polymers through a detailed understanding of polymer degradation chemistry. New analytical techniques continue to facilitate that analysis. Creative chemists are at work developing new approaches and new, more thermally stable organic structures. Mathematical fire models are becoming more sophisticated and more broadly applicable. Tests are becoming better understood and some provide data directly usable in fire models. And we are beginning to see the development of science-based regulations for furniture and building materials (particularly in Europe).

Table I. U.S. Plastics Production—1999 vs. 1998
(millions of pounds, dry weight basis[a]) (6)

Resin	U.S. Plastics Production	
	1999	% Chg 99/98
Epoxy	657	2.8
Urea	2,691	4.3
Melamine	294	1.4
Phenolic	4,388	11.4
Total Thermosets	8,030	7.8
LDPE	7,700	1.6
LLDPE	8,107	12.2
HDPE	13,864	7.3
PP	15,493	12.1
ABS	1,455	1.6
SAN	123	0.8
Other Styrenics	1,644	-0.6
PS	6,471	3.8
Nylon	1,349	5.0
PVC	14,912	2.8
Thermoplastic Polyester	4,846	9.6
Total Thermoplastics	75,964	6.7
Subtotal	83,994	6.8
Engineering Resins[b]	2,765	0.0
All Other Resins	10,702	4.3
Total Engineering & Other	13,467	3.4
Grand Total	97,461	6.3

[a]Except Phenolic resins (reported on a gross weight basis).

[b]Includes: Acetal, Granular Fluoropolymers, Polyamide-imide, Polycarbonate, Thermoplastic Polyester, Polyimide, Modified Polyphenylene Oxide, Polyphenylene Sulfide, Polysulfone, Polyetherimide and Liquid Crystal Polymers.

4

Table II. Resin Sales by Major Markets (millions of pounds) (7)

Major Market	1999	Compound Growth Rate 1995-1999
Transportation	3,836	3.2%
Packaging	21,270	5.6%
Building & Construction	19,072	8.9%
Electrical/Electronic	3,256	3.2%
Furniture & Furnishings	3,587	3.0%
Consumer & Institutional	11,802	7.2%
Industrial/Machinery	1,043	6.7%
Adhesives/Inks/Coatings	2,065	3.6%
All Other	10,446	8.9%
Exports	8,622	4.7%
Total Selected Plastics	84,999	6.4%

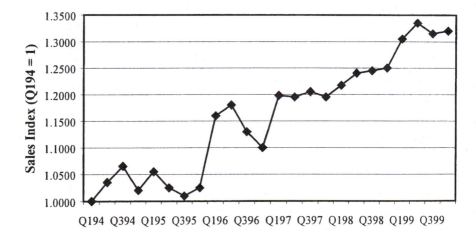

Figure 1. Six-Year Growth Graph: Trend of Sales and Captive Use
for Major Plastic Resins (8)

NOTE: Graph represents total sales and captive use of all thermoset and thermoplastic resins. The graph is an index, only meant to display performance of the industry, using 1994 as a base.

There are many diverse approaches to enhancing the fire stability of polymers; in the past the most common approach has involved the addition of additives. Ten years and more ago halogenated fire retardants, synergised by antimony oxide, were the method of choice to enhance the fire retardancy of many polymers. At this time there is a strong emphasis on non-halogenated fire retardants.

As one looks at previous *Fire and Polymers* volumes, topics have clearly changed. In 1990 fire toxicity was the first section with six papers (9). In 1995 there again was a section on fire toxicity with seven chapters (10). In this volume there is but one paper, Chapter 25. This represents the growing recognition that carbon monoxide is clearly the primary toxic gas in real fires, and its concentrations are more dependent upon the scenario than on the materials present. In 1990 there was a section on fire and cellulosics; again, in this volume there is only one chapter on wood, Chapter 28. In the 1995 volume there were twelve chapters on tests and regulation. In the current volume there are really only two. In the current volume the heart of current research is in the area of non-halogen approaches to flame retardancy, including additives and intrinsically fire-retardant polymers (fire-smart polymers). The topic of greatest interest is nanocomposites, although work has yet to get beyond decreases in heat release rates. With a considerable interest in "ecofriendly" materials, the need for new approaches to flame-retardant polymers has increased. That need is represented by half the papers in this volume. That is not to say, however, that very traditional materials are not recyclable; see Chapter 22. This peer-reviewed volume is designed to represent the state of the art.

References

1. Karter, Jr., M. J. *Fire Loss in the United States During 1999;* National Fire Protection Association: Quincy, MA, September 2000.
2. *The National Fire Problem;* United States Fire Administration National Fire Data Center; http://www.usfa.fema.gov/nfdc/national.htm.
3. *Residential Properties;* United States Fire Administration National Fire Data Center; http://www.usfa.fema.gov/nfdc/residential.htm.
4. *The Overall Fire Picture–1999;* United States Fire Association National Fire Data Center; http://www.usfa.fema.gov/nfdc/overall.htm.
5. Nelson, G. L. The Changing Nature of Fire Retardancy in Polymers. *Fire Retardancy of Polymeric Materials;* Grand A. F.; Wilkie, C. A.; Eds; Marcel Dekker: NY, 2000, pp 1-26.
6. APC Plastics Industry Producers' Statistics Group, as complied by Association Survey Resources, LLC. APC estimates. http://www.american plasticscouncil.org/benefits/economic/pips_99_sales_production.html.

7. American Plastics Council; http://www.americanplasticscouncil.org/ benefits/economic/pips_resin_markets_sales.html.
8. American Plastics Council; http://www.americanplasticscouncil.org/ benefits/ economic/pips_six_year_graph.html.
9. *Fire and Polymers, Hazards Identification and Prevention;* Nelson, G. L.; Ed.; American Chemical Society: Washington, DC, 1990.
10. *Fire and Polymers II, Materials and Tests for Hazard Prevention;* Nelson, G. L.; Ed.; American Chemical Society: Washington, DC, 1995.

Nanocomposites

One of the most exciting areas in polymer science is nanocomposites. The presence of a small amount of clay seems to enhance almost every property of a polymer, and fire retardancy is no exception. The initial paper (Morgan et al) is a contribution from the group at NIST which covers nylon and polyamide nanocomposites. This paper describes work which has been carried out as part of a NIST – industry consortium in the nanocomposite area. The results clearly show a significant reduction in the rate of heat release in the presence of the clay.

The second paper describes work in polystyrene nanocomposites using both clays and graphites as the nano material. While the rate of heat release is usually reduced in the presence of the clay, the materials do still burn and nanocomposites have not yet been discovered which will achieve a V-0 rating in the UL 94 test. This paper describes potential synergy between clays and conventional fire retardants.

Chapter 2

Flammability Properties of Polymer–Clay Nanocomposites: Polyamide-6 and Polypropylene Clay Nanocomposites

Alexander B. Morgan[1], Takashi Kashiwagi[1], Richard H. Harris[1], John R. Campbell[2], Koichi Shibayama[3], Koichiro Iwasa[3], and Jeffrey W. Gilman[1,*]

[1]Building and Fire Research Laboratory, National Institute of Standards and Technology, 100 Bureau Drive, MS 8652, Gaithersburg, MD 20899
[2]GE Corporate Research and Development, General Electric Company, Schenectady, NY 12301
[3]High Performance Plastics Headquarters, Sekisui Chemical Company, Ltd., Osaka, Japan 618-8589

Polymer layered-silicate (clay) nanocomposites have the unique combination of reduced flammability and improved physical properties. These results led to a NIST-industrial consortium which investigated the flammability properties of these nanocomposites further. A series of thermoplastic polymer nanocomposites were studied, along with modification of specific nanocomposite parameters to better understand their effect on the reduced flammability of the nanocomposite. The most important result from this work was the discovery that a clay-reinforced carbonaceous char forms during the combustion of the nanocomposites. This is particularly significant for systems whose base resin produces little or no char when burned alone. Further, we observed reductions in heat release rates of 70 % to 80 % for these thermoplastic polymer nanocomposites.

Introduction

Recent work performed at NIST showed that polymer layered-silicate (clay) nanocomposites exhibit reduced flammability and improved physical properties at low cost.(1,2,3,4) However, the details of the fire retardant mechanism are not well understood. In October of 1998 a NIST-industrial consortium was formed to study the flammability of these unique materials. The focus of research within this consortium was on the development of a fundamental understanding of the fire retardant (FR) mechanism of polymer clay nanocomposites. To better understand the FR mechanism, the affect of varying specific nano-structural parameters on the flammability properties of polymer clay nanocomposites was examined. Two of the polymer systems studied were, polypropylene (PP) and polyamide-6 (PA-6). The parameters which were modified were: 1) intercalated vs. delaminated (exfoliated) nanocomposites, 2) nanocomposites with different clay loading levels, and 3) the effect of a co-additive, specifically one which chars and could act synergistically with the nanocomposite.

Experimental[‡]

In general, the PP and PA-6 used to prepare the nanocomposites were compounded via melt blending in a twin-screw extruder.(5) PA-6 clay nanocomposites were also prepared via the *in situ* process developed by Unitika(6) and Toyota.(7) The *in situ* prepared PA-6 materials were obtained from UBE Inc.

Nanocomposite Preparation. PA-6 nanocomposites were prepared on a twin screw extruder (Welding Engineer, non-intermeshing, counter-rotating, die and 4 barrel segments at 246 °C, 41.9 rad/s [400 rpm], feed rate 6 kg/h). PA-6 (Capron C1250). To prevent segregation powder was mixed with organic treated montmorillonite (MMT) (Southern Clay Products, SCPX 1980, dimethyl-bis(hydrogenated tallow) ammonium MMT, d-spacing = 2.3 nm), and was then combined with PA-6 pellets in the extruder. Polyphenyleneoxide (PPO) was blended with two of the PA-6 formulations. PP-nanocomposites were

---

[‡] The policy of the National Institute of Standards and Technology (NIST) is to use metric units of measurement in all its publications, and to provide statements of uncertainty for all original measurements. In this document however, data from organizations outside NIST are shown, which may include measurements in non-metric units or measurements without uncertainty statements. The identification of any commercial product or trade name does not imply endorsement or recommendation by the National Institute of Standards and Technology.

compounded on a twin-screw extruder (JSW, TEX30 alpha, 32 mm screw diameter, co-rotating, L/D ratio: 51, zone temperatures: 170 °C to 190 °C, die temperature 190 °C, feed rate 15 kg/h). PP (Mitsubishi Chemical, Novatech EA9, melt flow index: 0.5 g/ 10 min, d = 0.9 g/cm^3) was compounded with organic treated MMT from Southern Clay Products (SCPX 1980, dimethyl-bis(hydrogenated tallow) ammonium MMT) and organic treated MMT from Nanocor (ODA Nanomer, octadecyl ammonium MMT). Polypropylene-graft-maleic anhydride (PP-g-MA, Sanyo Kasei, acid number: 26 mg KOH/g, mole fraction MA = 0.9 %, M_w[8]: 40000, Tg 154 °C) was used as a compatibilizer for some of the PP formulations.

Characterization. Each of the nanocomposite systems prepared was characterized using X-ray diffraction (XRD), and Transmission Electron Microscopy (TEM). TEM and XRD was done by the consortium member that prepared the sample. In regards to the uncertainty of XRD measurements performed at NIST, some comments about the use of XRD when analyzing polymer-clay nanocomposites are needed. The d-spacing uncertainty is ± 0.06 nm (one sigma). Material processing conditions, thermal stability of the organic treatment in the gallery, and the polymer itself can all affect the d-spacing. Even at larger spacing increases, one cannot say definitely how much polymer has entered into the gallery, thus defining the amount of intercalation or delamination which occurred during the synthesis of the polymer-clay nanocomposite. XRD is a useful screening tool, but the results provided by XRD cannot be used alone to define the exact nature of the nanocomposite. XRD only gives the distance between clay layers, thus revealing the relationship of the clay layers to themselves, but giving no information on clay-polymer interactions. XRD does not reveal how well dispersed the clay is throughout the polymer, nor does it define the degree of intercalation or exfoliation. Low magnification TEM will reveal how well dispersed the clay is throughout the polymer, and it can also show qualitatively the degree of intercalation and exfoliation which occurred. Of final note, XRD results do not describe clay dispersion accurately, and can be deceptive; the lack of a peak by XRD suggests an exfoliated nanocomposite, however it can also occur simply when the layers have become dissordered but not exfoliated or dispersed.(9)

Cone Calorimetry. All thermoplastic polymer nanocomposites were subjected to injection molding to give disks, (7.5 cm diameter × 0.8 cm thick) for the Cone Calorimeter tests. Cone Calorimeter experiments were performed at an incident heat flux of 50 kW/m^2.(10) Peak heat release rate (HRR), mass loss rate (MLR), specific extinction area (SEA), ignition time (t_{ign}), carbon monoxide yield, carbon dioxide yield, and specific heat of combustion data were gathered, and are generally reproducible to within ± 10 % when measured at 50 kW/m^2 flux. The cone data reported here are the average of three replicated experiments.

The specific uncertainties (one sigma) are shown as error bars on the plots of the Cone data. In many casses the uncertainty of the peak HRR is only ± 5 % .

Gasification. The specific design of the gasification device built at NIST has been reported previously.(11) The gasification device allows pyrolysis, in a nitrogen atmosphere, of samples identical to those used in the Cone Calorimeter, without complications from gas phase combustion, such as heat feedback and obscuration of the sample surface by the flame. The uncertainty in the measurement of interest in the gasification data is shown in each plot as an error bar.

Results and Discussion

PA-6 nanocomposites. Specific PA-6 nanocomposite formulations were made to allow ealuation of: (1) intercalated [polymer between clay layers, but primary clay particle stacks remain] versus delaminated nanocomposites [individual clay layers dispersed throughout polymer, no primary particle stacks remaining], (2) the effect of silicate loading level and (3) the effect of incorporating a charring-resin, polyphenyleneoxide (PPO), into a nanocomposite blend. The PA-6 nanocomposites were prepared with an intercalated nanomorphology by extruding the PA-6 with dimethyl, dihydrogenated-tallow ammonium treated MMT (SCPX1980). XRD analysis showed the interlayer spacing to be 2.45 nm, an increase of 0.15 nm. TEM (Figure 1) shows tactoids dispersed in the PA-6.

Preparation of intercalated PA-6 nanocomposites allows comparison to the delaminated PA-6 nanocomposite (UBE) which is prepared by the *in situ* method. The comparison of the heat release rate (HRR) behavior of the intercalated and the delaminated PA-6 nanocomposite is shown in Figure 2. The HRR curves are not significantly different. This indicates that intercalated and delaminated nanomorphologies are equally effective at reducing the flammability (HRR) of PA-6 nanocomposites made using MMT. However, a statistically significant difference in ignition times (t_{ign}) is evident between the intercalated and delaminated nanomorphologies from the HRR data in Figure 2. Specifically, the intercalated sample had an t_{ign} of 40 s compared to the t_{ign} of 80 s for the delaminated sample. The t_{ign} of the delaminated sample is similar to that for the pure PA-6 (t_{ign} 70 s). This shorter t_{ign} may be due to some physical effect (thermal conductivity, radiation absorption) or a chemical effect (thermal stability, volatile organic treatment).(12) In terms of possible chemical effects both the different methods of preparing the nanocomposites and the different MMT treatments may contribute to the earlier t_{ign}; the delaminated PA-6 nanocomposite sample is made via the *in situ* polymerization method, which uses an amino acid MMT treatment that becomes covalently bonded to the PA-6 as an end-group during the

*Figure 1. TEM of PA-6/5 % MMT (SCPX 1980) showing the intercalated
tactoids dispersed in the PA-6.*

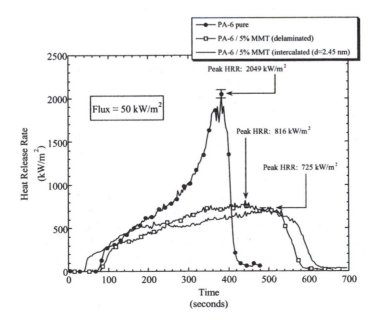

*Figure 2. Heat release rate (HRR) data for pure PA-6, and intercalated and
delaminated PA-6/MMT (mass fraction 5 %) nanocomposites.*

polymerization. The intercalated PA-6 nanocomposite, prepared via melt blending, at 246 °C, with a quaternary alkyl ammonium treated MMT, does not bond the MMT treatment to the polymer. This may reduce t_{ign}, since the decomposition temperature of the quaternary alkyl ammonium treated MMT (200 °C) is 100 °C lower than that for the delaminated PA-6 nanocomposite.(*1*) An additional complicating effect may be due to processing. Recently, using solid state NMR, we found evidence for decomposition of the both the *in situ* prepared (delaminated) PA-6 nanocomposite and the melt blended PA-6 nanocomposite during processing.(*13, 14*) The products generated, monomer and tertiary amine, during processing supply volatile fuel early in the burn and may counteract some of the flame retardant effect.

One could argue: the above comparison of intercalated and delaminated PA-6 nanocomposites is somewhat tenuous because of the different methods used to prepare the samples; however in a separate study, where the primary goal was to develop an NMR method to quantitatively characterize the degree of MMT delamination, the opportunity also arose to more clearly compare the effect of the degree of delamination on HRR. Three PA-6/MMT nanocomposites were prepared via melt blending PA-6 and the quaternary alkylammonium MMT SCPX1980. Three different levels of MMT delamination were prepared by varing the residence time in the twin-screw extruder. The samples were exaustively characterized using XRD and TEM. The peak HRR for the 2 samples with incomplete degrees of delamination are essentially identical; the peak HRR for the more delaminated sample is 100 kW/m^2 lower than the two samples with incomplete delamination. This is a statistically significant difference, and in contrast to the results discussed above it reveals that, as with most other properties, the flame retardant effect is maximized in fully delaminated nanocomposites. Full details of this study will be reported elsewhere.(*13*)

The effect of varying the MMT loading in PA-6 nanocomposites on the HRR is shown in Figure 3. The reduction in peak HRR improves as the mass fraction of MMT increases. The additional improvement for the PA-6/MMT nanocomposite with a MMT mass fraction 10 % is somewhat unusual when compared to the result for the other polymer nanocomposites we evaluated. (*15*) Usually there is little improvement above the 5 % loading level.

In our initial studies on the flammability of PA-6 nanocomposites we found that a layered-silicate carbonaceous residue formed during combustion.(*1*) However, there was very little (1 % to 2 %) <u>additional</u> carbonaceous char formed. We felt the use of an additive that would introduce additional carbonaceous char might enhance the flame retardant effect. To this end, PA-6 and polyphenylene oxide (PPO) were extruded with the organic modified MMT SCPX1980. The introduction of PPO into the PA-6/MMT nanocomposites gave very little improvement in HRR when added at the 5 % level. When 10 % PPO is added the HRR is lowered significantly, however, this is in part due to the inherent lower

HRR of PPO versus PA-6. The cone data for this sample showed that the PA-6/10 % MMT actually out performs even the PA-6/10 % PPO/5 % MMT. The char yield of PPO is 40 %, use of another polymer that has a higher char yield may be necessary to see the effect we envisioned.

In addition to measuring HRR the Cone Calorimeter also measures other fire-relevant properties such as mass loss rate (MLR), specific heat of combustion (H_c), and specific extinction area (SEA, a measure of smoke density), carbon monoxide yield, and carbon dioxide yield. The HRR the MLR data and the residue yield for the PA-6 nanocomposites discussed were the only parameters affected by the presence of nano-dispersed MMT. The MLR data follows the loss of fuel from the condensed phase into the gas phase. In this case the MLR follows the volatilization of PA-6 decomposition products (primarily caprolactam). Figure 4 shows the MLR data for the intercalated and delaminated PA-6/MMT (mass fraction 5 %) nanocomposites.

Comparison of Figure 2 to Figure 4, and recalling that none of the other parameters measured in the Cone calorimeter were strongly affected by the presence of nano-dispersed MMT, reveals that the nano-dispersed MMT primarily reduces the HRR by reducing the MLR (fuel feed rate) of the nanocomposite. This is consistent with the results we found in our initial studies of the delaminated PA-6/MMT.(*1*)

The above series of PA-6/MMT nanocomposites were also evaluated using the Gasification apparatus; the MLR data showed the same trends as observed in the Cone calorimeter experiments. Furthermore, we observed from the video data, that a black-residue formed on the sample surface at about 150 s into the gasification experiment. The formation of this residue coincided with the reduced MLR. An additional observation we made was that the mass loss initiates earlier for most of the PA-6/MMT nanocomposites as compared to the pure PA-6. This is analogous to the observed shorter t_{ign} for the melt-blended PA-6/MMT nanocomposites found in the HRR data.

PP nanocomposites. The focus of the experiments for the PP nanocomposites was to study the affect of varying the following nanocomposite parameters: (1) intercalated versus delaminated nanocomposites, and (2) nanocomposites with different silicate loading levels. Each PP/MMT nanocomposite prepared was characterized by XRD and TEM. The XRD data, and a summary of the nanomorphology, based on TEM and XRD, of each sample are shown in Table I. TEM shows that the PP/5 % MMT samples, made using either ODA Nanomer or SCPX1980, were intercalated nanocomposites with little or no delaminated layers (Figure 5). TEM of the PP compatibilized using PP-g-MA (PP/15 % PP-g-MA/MMT) revealed a mixed intercalated/delaminated nanomorphology, due to the presence of the PP-g-MA (Figure 6).

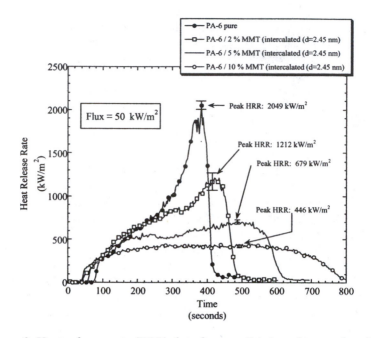

Figure 3. Heat release rate (HRR) data for pure PA-6, and intercalated PA-6/MMT nanocomposites (mass fractions 2 %, 5 %, and 10 %).

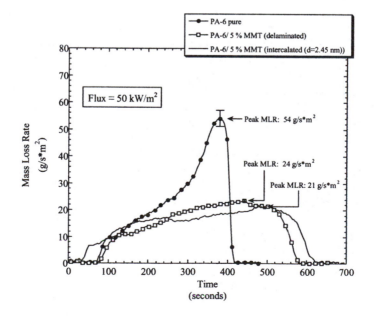

Figure 4. MLR data for PA-6 pure, intercalated and delaminated PA-6/MMT (mass fraction 5 %) nanocomposites.

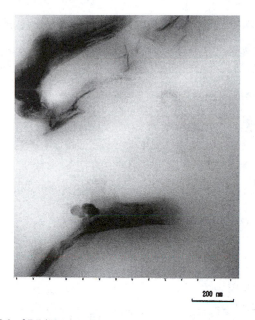

Figure 5. TEM of PP/5% MMT (SCPX 1980) showing intercalated tactoid structure.

Figure 6. TEM of PP/15% PP-g-MA/5% MMT (ODA Nanomer) showing delaminated/interlaminated nanomorphology.

Table I. XRD and TEM data for PP/MMT nanocomposites.

Sample	PP/MMT d-spacing (nm)	d-spacing change (nm)	TEM
PP/5 % MMT (SCPX1980)	2.63	0.09	*Int*
PP/15 % PP-g-MA/5 % MMT (SCPX1980)	3.68	1.14	*Int/Del*
PP/15 % PP-g-MA/2 % MMT (SCPX1980)	*	*	*Int/Del*
PP/15 % PP-g-MA/10 % MMT (SCPX1980)	3.22	0.68	*Int/Del*
PP/5 % MMT (ODA Nanomer)	2.71	0.86	*Int*
PP/15 % PP-g-MA/5 % MMT (ODA Nanomer)	3.56	1.71	*Int/Del*
PP/15 % PP-g-MA/2 % MMT (ODA Nanomer)	3.56	1.71	*Int/Del*
PP/15 % PP-g-MA/10 % MMT (ODA Nanomer)	3.50	1.65	*Int/Del*

Original d-spacing for: SCPX1980 = 2.54 nm, ODA Nanomer = 1.85 nm. *Int* = Intercalated. *Int/Del* = Intercalated/delaminated. * No XRD peak

The comparison of intercalated versus delaminated nano-morphologies, for PP nanocomposites, is somewhat difficult, since no completely delaminated sample was prepared. Yet, the HRR data shown in Figure 7 reveal that the mixed - intercalated and delaminated - sample (PP/15 % PP-g-MA/5 % MMT-ODA Nanomer) has a significantly lower HRR than the intercalated sample (PP/5 % MMT-ODA Nanomer) from 200 s to 400 s. This was also true for the PP/ SCPX1980 nanocomposites.

The HRR data for the PP nanocomposites with different MMT loading levels are shown in Figure 8. While the reduction in HRR is greater for the 5 % MMT sample compared to the 2 % MMT sample, only a small benefit is derived from increasing the loading from 5 % to 10 %. We have seen this in all of the nanocomposite systems we studied except for the PA-6 nanocomposites (Figure 3).

An additional observation, discernable from the HRR data in Figure 8, is that for the first 60 s of the burn the HRR for the nanocomposites is <u>higher</u> than that for the PP/PPgMA sample. This may be due to some physical effect (thermal difusivity, radiation absorption) or a chemical effect (thermal stability, volatile organic treatment). This is similar to the shorter ignition times (t_{ign}) we see in the

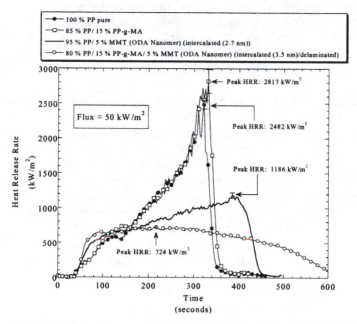

Figure 7. HRR plots for pure PP, PP/15% PP-g-MA, PP/5% MMT (intercalated) nanocomposite and PP/PP-g-MA/5% MMT (intercalated/delaminated) nanocomposite.

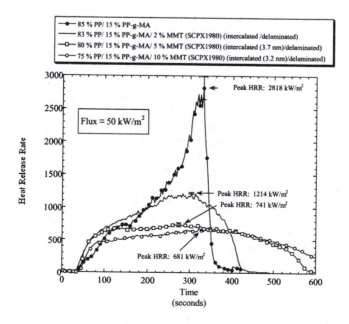

Figure 8. HRR plots for PP/15% PPgMA and PP/PPgMA/5% MMT (intercalated /delaminated) nanocomposite with 2%, 5%, and 10% MMT.

PA-6 nanocomposites (Figure 2). The use of additional char enhancing additives, or conventional flame retardants which delay ignition could counteract these effects. Indeed, recently published reports show that the use of intumescents, melamine, poly(tetrafluoroethylene) (PTFE), or red phosphorus combined with nanocomposites gives UL 94 V-0 ratings in a variety of polymer systems.(*16,17,18,19*)

The reduced MLRs for PP/MMT nanocomposites found in the gasification experiments closely follow the reductions in HRR from the Cone Calorimeter experiments discussed above. The formation of additional carbonaceous char was also seen in the PP nanocomposites. This is best observed in the digitized images from the videos of the gasification experiments (Figure 9). The critical role MMT and PPgMA play in forming carbonaceous char is evident by comparing the left image to the center image. Comparing the center and right images shows the effect of MMT loading on the volume and quality of the char formed. Here, both PPgMA and MMT must be present for additional carbonaceous char to be formed. The higher HRR behavior early in the burn, noted above, is also observed as higher MLR in the first part of the gasification experiments for the PP/MMT nanocomposite samples. Again, this may be from the physical or chemical effects discussed above.

Conclusions

The observed reductions in HRR for the PA-6 and PP nanocomposites are quite significant. Other than the greatly reduced HRR, the most important result from the Consortium's first year of work is the formation of a clay-reinforced carbonaceous char during combustion of nanocomposites. This is particularly significant since PP and PA-6 produces little or no char when burned alone. It appears from the gasification data (videos and mass loss data) that this clay-reinforced carbonaceous char is responsible for the reduced mass loss rates and hence the lower HRRs. Initially higher HRR and shorter t_{ign} are observed in many of the nanocomposites, and the origin of this effect needs to be better understood especially if UL94 V0 ratings are to be obtained for nanocomposites alone.

We conclude that delaminated nanocomposites perform better than intercalated nanocomposites in PP and PA-6 nanocomposites. In terms of the effect of loading level, the effectiveness of the nanocomposite approach to reducing HRR, in most cases, levels off at 5 % silicate loading, but some additional improvement is seen at 10 % loading for intercalated PA-6/MMT. And finally, the use of a char-enhancer (PPO) did not decrease the flammability of the PA-6 nanocomposites, but other char-enhancing co-additives should be explored.

Figure 9. Digitized photos of gasification residues from PP/5 % MMT (SCPX1980) (left), PP/15 % PPgMA/2 % MMT (SCPX1980) (center), and PP/15 % PPgMA/5 % MMT (SCPX1980) (right).

22

Acknowledgements

The authors wish to thank the following members of NIST-MSEL: Dr. Alan Nakatani for assistance in use of the micro twin-screw extruder, Dr. Catheryn Jackson for TEM assistance, and Dr. Terrell Vanderah for use of XRD facilities. Finally, we wish to thank the following organizations for funding of this work: Air Force Office of Scientific Research, Federal Aviation Administration (DTFA 03-99-X-9009) and the Flammability of Polymer-Clay Nanocomposites Consortium.

References

1. Gilman, J. W.; Kashiwagi, T.; Lichtenhan, J. D. *SAMPE Journal*, **1997,** *33,* 40-46.
2. Gilman, J. W.; Kashiwagi, T.; Lomakin, S.; Giannelis, E.; Manias, E.; Lichtenhan, J.; Jones, P. In *Fire Retardancy of Polymers: the Use of Intumescence*; LeBras, M., Camino, G., Bourbigot, S., and Delobel, R. The Royal Society of Chemistry: Cambridge, 1998: pp 203-221.
3. Gilman, J. W. *App. Clay. Sci.* **1999,** *15,* 31-49.
4. Gilman, J. W.; Jackson, C. L.; Morgan, A. B.; Harris, R. H.; Manias, E.; Giannelis, E. P.; Wuthenow, M.; Hilton, D.; Phillips, S. *Chem. Mater.* **2000,** *12,* 1866-1873.
5. Giannelis, E., *Advanced Materials*, **1996,** *8,* 29-35.
6. Fujiwara, S., Sakamoto, T., Kokai Patent Application, no. SHO 51(1976)-109998, **1976.**
7. Usuki, A., Kojima, Y., Kawasumi, M., Okada, A., Fukushima, Y., Kurauchi, T. and Kamigaito, O., *J. Mater. Res.* **1993,** *8,* 1179-1184.
8. According to ISO 31-8, the term "Molecular Weight" has been replaced by "Relative Molecular Mass", symbol Mr. Thus, if this nomenclature and notation were used here, Mr,n instead of the historically conventional Mn for the average molecular weight (with similar notation for Mw, Mz, Mv) would be used. It would be called the "Number Average Relative Molecular Mass". The conventional notations, rather than the ISO notations, have been employed here.
9. Morgan, A. B.; Gilman, J. W.; Jackson, C. L. *Polym. Mater. Sci. Eng. (ACS),* **2000,** *82,* 270-271.
10. Babrauskas, V., Peacock, R. D. *Fire Safety Journal* **1992,** *18,* 255-262.
11. Austin, P. J.., Buch, R. B., Kashiwagi, T. *Fire and Mater.* **1998,** 22, 221.
12. The Cone data for pure PA-6 shown in the HRR and MLR plots is for the material from UBE. However the pure PA-6 used by GE to prepare the

nanocomposites (from Allied Signal) has very similar flammability properties to the UBE material.

13. Vanderhart, D.; Asano, U.; Gilman, J. W. *Macromolecules* submitted.
14. Davis, R., VanderHart, D. L., Asano, A., and Gilman, J. W. manuscript in preparation.
15. Gilman, J. W.; Kashiwagi, T.; Morgan, A. B.; Harris, R. H.; Brassell, L. D.; VanLandingham, M.; Jackson, C. L. *National Institute of Standards and Technology Interagency Report, NISTIR 6531*, July 2000.
16. Bourbigot, S., LeBras, M., Dabrowski, Gilman, J., and Kashiwagi, T. Proceedings of 10th BCC Annual Conference, May 1999.
17. Inoue, H.; Hosokawa, T., Japan Patent, Jpn. Kokai tokkyo koho JP 10 81,510 (98 81,510), 1998.
18. Takekoshi, T. et al., US patent 5,773,502, 1998.
19. Klatt, M. et al., PCT Int. Appl. WO98 36,022, 1998.

Chapter 3

Recent Studies on Thermal Stability and Flame Retardancy of Polystyrene-Nanocomposites

Jin Zhu, Fawn Uhl, and Charles A. Wilkie

Department of Chemistry, Marquette University, P.O. Box 1881, Milwaukee, WI 53201-1881

Nanocomposites of polystyrene with both organically-modified clays and with graphite have been prepared. For both of these materials, which will impart nanometer dimension to the polymer, the thermal stability of the polymer, as measured by thermogravimetric analysis and Cone calorimetry, is enhanced. A synergistic combination of a clay nanocomposite with resorcinol diphosphate has also been studied.

Nanocomposites, based on layered inorganics, exhibit new and improved properties due to their nanometer dimension. They show increased stiffness and strength and enhanced thermal stability without sacrificing impact resistance. A recent report also indicates that nanocomposites may have improved flame retardancy relative to the pure polymers (1,2).

Currently, two approaches have been used for the formation of nanocomposites: blending and in-situ polymerization. Melt-blending is based on melt intercalation of the polymer and involves annealing a mixture of polymer and clay above the glass transition temperature, T_g, of the polymer(3,4).

Solution blending has also been practiced. In-situ polymerization is based on polymerization of monomers in the presence of clay (5,6,7,8). Since small molecules can easily insert into the galleries of the clay, in-situ polymerization can produce well-dispersed materials.

In the present communication, we review our recent studies on the thermal properties and flammability of polystyrene (PS) nanocomposites, using both clays and graphite as the material to give nanostructure to the polymer. These nanocomposites were prepared by in-situ polymerization(9,10,11). Several different organic modifications to the clay have been made in order to examine a variety of materials. In addition to the work on pure nanocomposites, we also describe in this paper a putative synergistic combination between a nanocomposite and a conventional fire retardant, resorcinol diphosphate, RDP, a product of Akzo-Nobel. In the case of graphite nanocomposites, we have utilized the intercalation compound of potassium in graphite, KC_8 (12, 13).

Polystyrene-Clay Nanocomposites

Several commercial organically modified clays as well as some modified clays that were prepared in these laboratories have been used. There is a clear difference amongst the various organic modifications in terms of the intercalated or exfoliated nature of the nanocomposites. Most preparations of PS nanocomposites have been produced through the melt blending process and the typical result is that some mixture of intercalated and exfoliated structure results. The normal method to obtain this information is the combination of powder x-ray diffraction, XRD, and transmission electron microscopy, TEM. The expansion of the d-spacing in the XRD measurement gives a firm indication that something has entered the gallery space and expanded the lattice. If this process continues to the point where there is no longer any registry between the clay layers, the XRD peak will disappear and the material is described as exfoliated (another term used for this is delaminated). The observation of an expanded d-spacing is a good indicator that an intercalated nanocomposite has been formed but the absence of an XRD peak does not necessarily mean that exfoliation has occurred, since the peak could be absent for experimental reasons. It is for this reason that TEM measurements are required to describe the state of the nanocomposite. Nanocomposites have been reported to offer barrier properties, to strengthen the polymer and to enhance the thermal stability of polymers. It is believed that the extent of exfoliation is extremely important in the first two examples while it appears to be much less important for fire retardancy.

The naturally occurring clay has a sodium counter ion and this is, of course, surrounded by water of hydration. This clay is not compatible with an organic polymer and these must be modified to make them organophilic. The normal process that is used is to ion-exchange the sodium with an organic ammonium salt, or some similar material. The commercial clays that have been studied contain some combination of methyl, long chain alkyl, typically C_{16} or C_{18}, and benzyl. In this laboratory we have prepared three salts, two are ammonium salts and both contain two methyls, one C_{16} chain, and the fourth substituent is a benzyl group attached to the ammonium nitrogen with a *para* vinyl(VB16) or a *para* benzyl alcohol (OH16). The third salt is a phosphonium salt, this is substituted with three phenyls and a C_{16}(P16). The XRD of all of these materials, except VB16, shows the expanded d-spacing which is typical of a nanocomposite. The TEM measurement confirm that all of these materials contain some mixture of intercalated and exfoliated structures. The singular exception, VB16, shows a completely exfoliated structure. It must be noted that the ammonium salt which is used for this material contains a styrene monomer unit and that polymerization may occur onto the clay counter ion. Thus in this system there will be both polymer grafted onto the clay as well as bulk polymer.

Thermogravimetric analysis studies on these nanocomposites show that the onset temperature of the degradation is increased by at least 50°C compared to virgin polystyrene. This very large increase indicates the substantially enhanced thermal stability of the system; this is also shown by the reduction in peak heat release rate in a Cone calorimetry experiment. For nanocomposites in which the mass fraction of clay is 3%, the average reduction in the peak rate of heat release at a flux of 35 kW/m^2 is 57% for all systems which have been examined. There is a small reduction in the peak heat release rate as the mass fraction of clay increases from 0.1% to 10% but the average reduction is 48% over this large concentration range.

The greatly reduced rate of heat release has been reported frequently in the literature and the usual mechanism which is cited to explain this is that the clay acts as a barrier and prevents both mass transport of the polymer degradation products from the bulk and heat transfer into the bulk. It has not been widely reported that there is still not an example of a nanocomposite which will pass the Underwriter's Laboratory fire protocol, UL-94. In this test a thin piece of the polymer is ignited and one measures the ease of extinction of the flame. The standard which is normally required to offer a commercial flame retardant is the V-0 rating which means each sample self-extinguishes within 10 seconds; there are some additional requirements for this rating. In order to attempt to improve on this, we have attempted to find an appropriate synergistic combination of a conventional flame retardant which can be combined with the clay to yield a suitable efficacious system.

Combination of Polystyrene-Clay Nanocomposite and Resorcinol Diphosphate

The Akzo-Nobel Company offers resorcinol diphosphate, RDP, as a flame retardant which is claimed to have activity in both the condensed and vapor phases; this seemed a suitable starting point for our investigation since it is not known what type of activity the optimum flame retardant should have for synergy with a clay.   RDP has been investigated as a flame retardant for PC\ABS blends as well as other systems (14), but it, as well as other phosphate fire retardants is not suitable in styrenics, since the level which must be used will cause plasticization of the polymer. This was confirmed in this study, and it was also found that the presence of the clay counteracts the plasticization effect from the phosphate and the system could be useable.

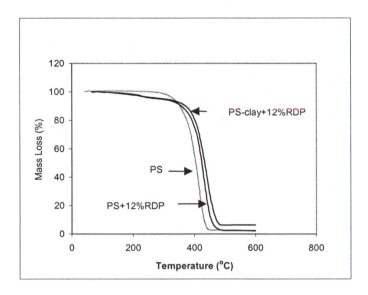

Figure 1. TGA curves for polystyrene, polystyrene + RDP and PS-clay nanocomposite + RDP.

The combination of the nanocomposite with RDP has been prepared both by bulk polymerization, following a protocol identical to that which has been previously published with the addition of the appropriate amount of RDP to the styrene, and by merely blending using a Brabender mixer. The materials which result from this have been analyzed by thermogravimetric analysis and by Cone calorimetry. In Figure 1 are presented the TGA curves for polystyrene, and this polymer combined with RDP and with both RDP and a clay. One can see that the degradation commences at an earlier temperature in the presence of RDP, no doubt due to the volatilization of the RDP, but after some degradation has occurred, the combinations with RDP are more thermally stable. The amount of RDP has no effect on the TGA curves, as shown in Figure 2. The TGA data, temperature at which 10% and 50% mass loss occurs and the fraction of non-

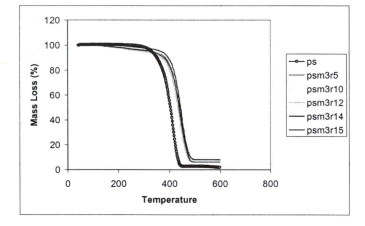

Figure 2. TGA curves of polystyrene-clay nanocomposites which also contain varying amounts of RDP; the number at the end of the legend is the mass fraction of RDP. The curves for the combination of RDP and the nanocomposite overlap and are consistently higher than that of vrtgin polystyrene.

Table 1. TGA results of PS nanocomposites, 3% mass fraction of clay, containing varying amounts of RDP.

Composites	$T_{10\%}(^oC)$	$T_{50\%}(^oC)$	Non-volatiles at 600°C
PS	351	404	0
Nanocomposite+5%RDP	386	440	8
Nanocomposite+10%RDP	381	436	9
Nanocomposite +12%RDP	370	435	7
Nanocomposite +14%RDP	378	438	6
Nanocomposite +15%RDP	396	444	8
PS+5%RDP	360	410	3
PS+10%RDP	371	423	2
PS+12%RDP	358	424	2
PS+14%RDP	348	415	4
PS+15%RDP	363	427	4

volatile material which remains at 600°C, for PS with RDP and the PS-clay nanocomposite with RDP is shown in Table 1. One can see a rather consistent difference between PS + RDP and the PS-clay nanocomposite + RDP. The temperature at which 10% degradation has occurred is consistently lower in the absence of the clay than in its presence.

Cone calorimetry has been carried out on the RDP-containing nanocomposites. It is not possible to mold polystyrene with RDP into Cone plaques, since the presence of the phosphate causes a great deal of plasticization of the polystyrene. Table 2 contains the Cone results and one can see that in all cases it is easier to ignite the RDP-containing nanocomposite than for virgin PS. Likewise, the peak rate of heat release, peak HRR, is decreased by an average of 66%, from a value of 1024 kW/m^2 in virgin PS to the average value of 345 kW/m^2 for the RDP-containing nanocomposites; the value for the nanocomposite without RDP is 449 kW/m^2. The time to burnout is increased by about 20% and the energy that is released up to the time at which all of the polystyrene has burned is decreased by 49%. It is somewhat surprising to observe that the percentage of mass which is lost up to the time at which all of the polystyrene has burned is decreased by only 12%. Since the amount of mass that is lost does not change by a large amount but the energy released does change by a substantial number, this must mean that the mass that is lost is different in the virgin PS compared to the nanocomposite plus RDP. The rate at which mass is lost is diminished in the presence of the clay and RDP and this may be related to the different materials lost comparing the two samples.

Table 2. Cone calorimetry results for polystyrene-clay nanocomposites, containing 3% of a dimethylbenzyl-C18 ammonium salt and varying amounts of RDP.

Materials	PS	Nano + 5%RDP	Nano + 12%RDP	Nano + 15%RDP
Time to ignition, s	35	15	15	15
PHRR, kW/m^2	1024	414	310	325
Time to PHRR, s	165	130	160	155
Time to burnout, s	190	232	227	234
Energy released through 190 s, kW	981	550	479	483
Average mass loss rate, mg/s	127	88	82	89
Mass loss at 190 s, %	86	79	67	79
Specific extinction area, m^2/kg	1572	1311	1443	1350

A graphical presentation of the peak HRR for virgin PS along with three RDP-containing nanocomposites with differing amounts of RDP is shown in Figure 3. One can easily see the tremendous difference between virgin PS and the nanocomposites which contain RDP. It is also clear that there is some small difference between the 5% RDP sample and those with larger amounts of RDP.

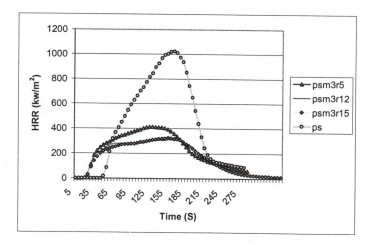

Figure 3. Peak heat release rates for polystyrene, PS, and nanocomposites containing 3% clay and either 5%, 12% or 15% RDP.

A graphical representation of the mass loss data is shown in Figure 4. Clearly there is some interaction between the polymer, clay, and phosphorus material. low temperatures there is increased mass loss from the nanocomposite plus RDP, a reflection of the volatility of the RDP, at about 170°C the curves overlap and above this temperature virgin PS looses mass at a more rapid rate.

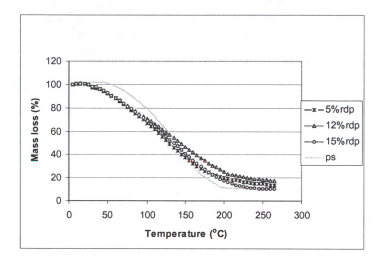

Figure 4. Mass loss data for nanocomposites containing RDP.

Polystyrene –Graphite Nanocomposites.

Polymer – graphite nanocomposites have been prepared using potassium graphite, KC_8 as the initiator. This intercalation compound contains atomic potassium inserted between graphite layers so anionic polymerization must occur between the graphite layers. The d-spacing increases from 3.35Å in graphite to 5.41Å in KC_8 and to 7.50 to 13.7Å in the polystyrene nanocomposite.

Thermogravimetric analysis shows an overall enhancement in thermal stability of the graphite nanocomposites relative to virgin polystyrene. Typically the temperature at which 10% degradation occurs increases from the normal value of 350°C for virgin polystyrene to a temperature above 400°C for the nanocomposite. In addition a significant amount of char, far exceeding the amount of graphite which has been added to the system, is produced.

Preliminary results from Cone calorimetry have shown that the peak heat release rate is decreased but the decrease is not as large as that observed for polystyrene-clay nanocomposites. The average reduction is 40% for the graphite nanocomposites but closer to 60% for the clay nanocomposites.

32

Conclusion

Polystyrene- nanocomposites in which the material with the nanometer dimensional character is either an organically-modified clay of graphite have been prepared and characterized. In both cases there is an enhancement in thermal stability as shown by TGA measurements and Cone calorimetry. Studies have been carried out in which the polystyrene-clay nanocomposite has been combined with a commercial flame retardant, resorcinol diphosphate, and synergy between the components is noted.

Acknowledgement

This work has been partially supported by the US Department of Commerce, National Institute of Standards and Technology, grant number 60NANB6D0119.

References

1. Gilman, J.W.; Kashiwagi, T.; Giannelis, E.P.; Manias, E.; Lomakin, S.; Lichtenhan, J.D.; Jones, P. in *Fire Retardancy of Polymers The Use of Intumescence*, Le Bras, M.; Camino, G.; Bourbigot, S.; and Delobel, R. Eds.; Royal Soc of Chem.: Cambridge, **1998**, pp. 203-221.
2. Gilman, J.W.; Kashiwagi, T.; Nyden, M.; Brown, J.E.T.; Jackson, C. L.; Lomakin, S.; Giannelis, E.P.; Manias, E. in *Chemistry and Technology of Polymer Additives*, Al-Maliaka, S.; Golovoy, A.; Wilkie, C. A.; Eds.; Blackwell Scientific: London, 1999, pp. 249-265.
3. Vaia, R.A.; Ishii, H.; Giannelis, E.P. *Chem. Mater.*, **1993**, 5, 1064-1066.
4. Burnside, S.D.; Giannelis, E.P. *Chem. Mater.*, **1995**, 7, 1597-1600.
5. Messersmith, P.B.; Giannelis, E.P. *J. Polym. Sci. Pt. A, Polym. Chem.*, **1995**, 33, 1047-1057.
6. Usuki, A.; Kojima, Y.; Kawasumi, M.; Okada, A.; Fukushima, Y.; Kurauchi, T.; Kamigaito, O., *J. Mater. Sci.*, **1993**, 8, 1179-1184.

7. Jao Goo Doh; Iwham Cho. *Polym. Bull.*, **1998**, 41, 511-518.
8. Akelah,A.; Moet A. *J. Mater. Sci.*, **1996**, 31, 3589-3596.
9. Zhu, J.; Wilkie, C. A.. *Polym. Intern.*, in press.
10. Zhu, J., Morgan, A. B., Lamelas, F., and Wilkie, C. A. manuscript in preparation.
11. Uhl, F., and Wilkie, C. A. manuscript in preparation.
12. Rüdorff, W. *Advan. Inorg. Radiochem.*, **1959**, 1, 223-266.
13. Hennig, GR. *Prog. Inorg. Chem.*, **1959**, 1, 125-205.
14. Levchik, S., Dashevsky, S., and Bright, D. A., 11th Annual BCC Conference on Recent Advances in Fire Retardancy of Polymers, Stamford, CT, 2000.

Fire Smart Polymers

Fire retardancy currently depends largely upon additive technologies in order to render polymers less flammable. In the future, we will increasingly see new polymeric systems in which the material is inherently less likely to undergo thermal degradation, *i.e.*, the polymer is fire smart. The first paper, by Pearce, Weil and Barinov, looks back over two decades of work on the utilization of polymers in which functional groups have been used to form more fire-resistant entities under fire conditions.

The paper by Williams et al. describes recent work in polyimides—materials that have the potential for use in aerospace applications. The last paper in the section moves into the area of fibers and describes poly(benzobisoxazole)s, a material that shows truly excellent performance in the cone calorimeter. The degradation of these fibers in nitrogen does not begin until 700 °C.

Chapter 4

Fire Smart Polymers

Eli M. Pearce, Edward D. Weil, and Victor Y. Barinov

Department of Chemical Engineering, Chemistry and Material Science,
Polytechnic University, 6 Metrotech Center, Brooklyn, NY 11201

This review paper surveys two decades of research in which functional groups on polymers have been utilized to form more fire-resistant structures under fire-exposure conditions. Examples are polyamides which convert to polybenzoxazoles, nitrile polymers with latent catalysts for triazine ring formation, styrylpyridine polymers which form stable cyclic crosslinks, chloromethylstyrene copolymers with latent Friedel-Crafts crosslinking, and "cardo" (looped) polymers as exemplified by phenolphthalein-containing epoxy resins which crosslink *via* their lactone rings under strong heating. Other examples are silanol oligomers which can impart pre-ceramic properties to hydrogen-bond-accepting polymers, and halogenated phenolic resins with enhanced flame retardant properties.

A promising approach towards reducing the flammability of polymer systems is to alter the condensed phase chemistry at elevated temperatures. Structure modification can alter the decomposition chemistry favoring the transformation of the polymer to a char residue. This can be achieved by the addition of additives that catalyze char rather than flammable products or by designing polymer structures that favor char formation (*1*).

This paper reviews the basic principles of our work in "smart polymer" flame retardancy and the relationship of polymer flammability to polymer structure with examples illustrating these concepts.

The theme of our research has been to find ways to obtain increased char formation by finding means for encouraging thermally induced stable crosslinking structures.

Aromatic Polyamides and Substituted Aromatic Polyamides

Various wholly aromatic polyamides (aramides) based on m- and p-phenylenediamines and isophthaloyl and terephthaloyl chloride have been synthesized and their properties and oxygen index values have been studied. Specific halogen-substituted aramides ortho-substituted on the diamine ring by displaceable groups have given substantial increases in char formation due to the formation, in part, of thermally stable benzoxazole units (Figure 1)(2).

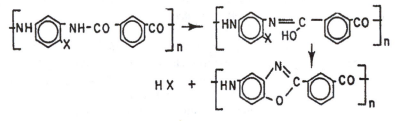

Figure 1. Thermal formation of benzoxazole units.
X represents halogen, nitro or cyano.

The effects of different substituents on the aromatic ring of the diamine have been explored by comparing their DTA and TGA behavior, their relative char yields at 700°C, and their oxygen indexes. The halogen, nitro and cyano-substituted polyamides have been found to produce the highest char yields. The high char yields are probably associated with crosslinking occurring at high temperatures. Attempts at correlating char yield with oxygen index indicated enhancement for the chloro-substituted aramides (3-5).

In order to understand these systems, studies were initiated on degradation and degradation products obtained from poly(1,3-phenylene isophthalamide) and poly(4-chloro-1,3-phenylene isophthalamide) and their model compounds. Their flammabilities were measured by the oxygen index method. The chlorophenylene polyamide had greatly reduced flammability as shown by a 10-15 higher oxygen index. Analysis of the chars of the two polymers at 700°C by TGA and elemental analysis showed that the chlorine caused a significant increase in the pyrolysis residue. Based on these results, we have suggested that the chlorine imparts flame retardancy by a combination of vapor- and condensed-phase mechanisms (5).

Our studies on substituted aramides were examples in which high temperature thermally stable aromatic heterocyclic rings were formed. It was found that halogen substitution affected significantly the thermal characteristics and flame resistance of poly(1,4-phenylene terephthalamide).

In the case of the halogenated polyamides the char yield enhancement and the flame resistance improvement are associated with halogen displacement and ring-forming reactions during their pyrolysis (6-10). It has been found that polyamides containing para polymeric linkages are more thermally stable then those containing meta linkages. We have investigated orthohalogen substituted aramides and have shown that at elevated temperatures benzoxazole structures were formed. The amount of char residue that was formed at these elevated temperatures was dependent on the nature and position of the halogen groups (9,10).

Various aramides which contain a nitro group on the amino substituted ring were synthesized and their thermal properties and degradation mechanisms were studied. Thermal decomposition of ortho-substituted aramides proceeds via a two-stage mechanism, whereas unsubstituted aramides decompose in one step. The first step represents the loss of HNO_2 and the second step is due to degradation of the resulting benzoxazole polymer (11).

Another series of substituted aramides have a cyano group on the arylamine ring. These also appear to undergo the benzoxazole ring formation if the cyano group is ortho to the amino group (12). A greatly increased char yield was found by thermogravimetric analysis for the ortho-cyano structure compared to the other positional isomers.

The flame resistance of polymeric materials can be enhanced with the modification of chemical structure together with the incorporation of additives such as catalysts. Among the investigated additives zinc chloride (a presumed Lewis acid catalyst) was shown to improve the char yields and flame resistance of the poly(1,3-phenylene isophthalamides) (13).

Vinylbenzyl Chloride-Styrene Copolymers and Friedel-Crafts Crosslinking

Polystyrenes tend to give copious volatile fuel and little char when exposed to fire temperatures. To overcome this behavior, we copolymerized a latent crosslinking component, vinylbenzyl chloride, with styrene at a range of monomer ratios (14). Antimony oxide or zinc oxide were found to be latent Friedel-Crafts catalysts. The char yield increased more than linearly with vinylbenzyl chloride content, as shown in Figure 2. The oxygen index also went up sharply with vinylbenzyl chloride content, as shown in the same figure.

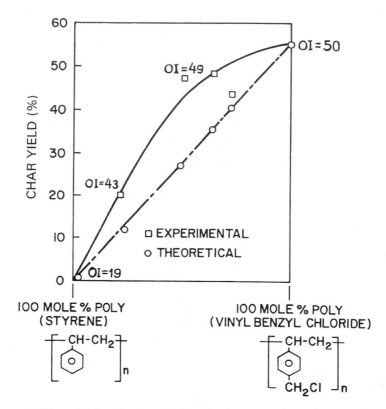

Figure 2. Char yield and oxygen index of polymers and copolymers of styrene and vinylbenzyl chloride

It should be noted that an alternative way to accomplish the same kind of Friedel-Crafts crosslinking in a polystyrene is to add a bifunctional or polyfunctional alkylating agent, rather than to put the alkylating group on the polymer. Early examples of this approach are by Brauman (*15*) and by researchers at Ciba-Geigy (*16*).

Nitrile Polymers

Styrene-acrylonitrile copolymers are strong, rigid, and transparent, they have excellent dimensional stability, high craze resistance, and good solvent resistance, but their flammability requires retardation for designed end uses. The effect of zinc chloride on the thermal stability of styrene acrylonitrile copolymers was studied. Our work showed that upon the addition of zinc chloride to styrene acrylonitrile copolymers, the initial thermal stability was decreased but char yield was significantly increased. A high temperature FTIR study showed that zinc chloride complexed with the nitrile group, and this, in turn, induced a modified degradation mechanism leading to thermally stable triazine ring formation which crosslinked the main chains (17,18).

A further example of a nitrile polymer which showed excellent char yield in the presence of Lewis acid catalyst is a cyanophenoxy-substituted phosphazene polymer, which appears to crosslink by nitrile group trimerization (19). When the trifluoroethoxy cyanophenoxy phosphazene is provided with a crosslinking catalyst and subjected to TGA in comparison to the uncatalyzed polymer, the char yield is increased several-fold (19).

Styrylpyridine Polymers

Another system in which ring formation accounts for increased char formation were epoxy resins and polyesters which contained styrylpyridine units (Figure 3).

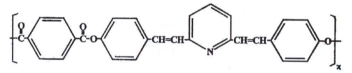

Figure 3. Styrylpyridine unit (in linear polyester)

In this case a Diels Alder addition reaction could account for these results. This gave both a crosslinking and ring formation reaction. An intermolecular Diels Alder reaction, involving the ethylene group as dienophile and (in a separate polymer chain) the ethylene group plus the adjacent 2,3-unsaturation of the pyridine ring as diene was proposed as the thermal crosslinking reaction (20). A subsequent Claisen-Cope rearrangement restores aromaticity to the pyridine ring and prevents reversal of the Diels Alder reaction, thus creating an extremely stable crosslink.

Styrylpyridine-containing polyesters are useful candidates for high temperature photoresists, because of the excellent thermal stability conferred by the rigid styrylpyridinium repeat units and the propensity to crosslink rather than

42

undergo intramolecular decomposition (*21*). This same property can be expected to favor flame retardancy.

"Cardopolymers"

Macromolecules of "cardopolymers" have loops, i.e., where the ring and the polymer backbone have one atom in common. A polymer having high aromaticity and/or pendant ring structure in the chain backbone could give high heat, thermal, and flame resistance. "Cardo"-type polycarbonates, polyesters and epoxy resins were prepared using either phenolphthalein or fluorenebisphenol to provide the "loops." A comparison of the usual bisphenol-A-based polycarbonate to the phenolphthalein-based polycarbonate is shown in Figure 4.

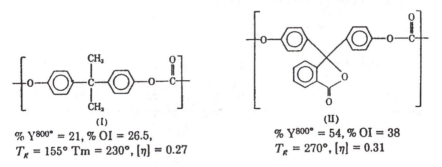

(I)
% Y$^{800°}$ = 21, % OI = 26.5,
T_g = 155° Tm = 230°, [η] = 0.27

(II)
% Y$^{800°}$ = 54, % OI = 38
T_g = 270°, [η] = 0.31

Figure 4. Comparison of properties of bisphenol-A polycarbonate and phenolphthalein polycarbonate

The glass transition temperatures (T_g) may be attributed to an increase of aromatic ring content in the chain backbone. The increase in char yield (%Y$^{800°}$) and increase in oxygen index (OI) can be attributed to the increased degree of aromaticity and the presence of the "cardo" structures (*22-26*).

The increase in flame retardancy and char yield was correlated with the thermal formation of ester crosslinks, demonstrable by a decrease in the lactone carbonyl and an increase in open-chain ester carbonyl in the infrared spectrum upon strong heating (*25-26*).

In a more recent study on "cardo" polymers (*27*), we incorporated phenolphthalein as a chain-extender into epoxy resins. The resin was then cured by a conventional crosslinker (dicyandiamide). The introduction of phenolphthalein provided us with a "cardo" polymer of which the loop was the lactone ring of the phenolphthalein. We found that this structural change increased the onset weight loss temperature (from 352°C without the ring to 393°C with the ring) and increased the TGA residue yield slightly (from 13% to

16% respectively) but did not decrease the flammability as measured by oxygen index (21.6% without phenolphthalein, 21.4% with phenolphthalein). No rating was obtained by the UL 94 test. We therefore took the additional step of introducing a phosphate, tetraphenyl resorcinol diphosphate (RDP). The commercial flame retardant made by Akzo Nobel is an oligomeric mixture. The largest component is the diphosphate (Figure 5).

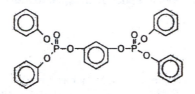

Figure 5. Tetraphenyl resorcinol diphosphate

With this phosphate added, better flame retardancy results were obtained with the epoxy chain extended with phenolphthalein than without the phenolphthalein, as shown in Figure 6.

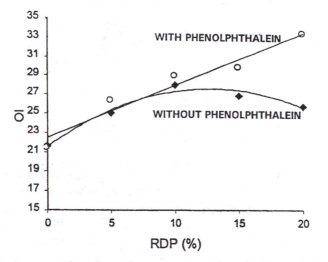

Figure 6. Oxygen index of cured epoxy resin with and without chain extension by phenolphthalein, at various loadings of tetraphenyl resorcinol diphosphate.

The infrared spectra of the dicyandiamide-cured phenolphthalein-chain-extended epoxy resin under strong heating to 350°C showed diminishment of the lactone carbonyl band (1760 cm^{-1}) of the phenolphthalein unit, and formation of probable amide groups at 1658 cm^{-1}. This crosslinking chemistry

44

was attributed to reaction of the lactone rings with residual NH groups left from the dicyandiamide cure.

The significant advantage of having the crosslinking site located on the polymer backbone instead of having the same functionality located in an additive was shown in this study by the poor flame retardant behavior of similar epoxy resins with the phenolphthalein unit present not in the backbone but in a diphosphate additive (tetraphenyl phenolphthalein diphosphate)(27).

Siloxane-Based Polymers

Poly(methylhydroxysiloxane) (PHMS) has been synthesized by selective oxidation of poly(methylhydrosiloxane)(PMHS) with dimethyldioxirane solution in acetone (Figure 7).

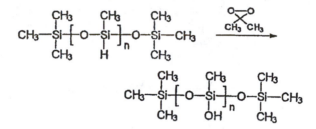

Figure 7. Preparation of poly(methylhydroxysiloxane)

These functional polysiloxanes gave ceramic-like materials by self-condensation at about 700°C under nitrogen atmosphere. The high yield of residue under N_2 implies a high degree of flame resistance. Miscibility studies with a variety of polymers showed that organic-inorganic hybrids could be formed for poly(vinylpyrrolidone), poly(vinylpyridine) and others having hydrogen-bond acceptor structures. Polysiloxanes with multiple silanol groups can be considered as useful components for blending with hydrogen-bond accepting polymers to impart fire-resistant and/or preceramic properties (28).

Phenolics

A variety of ring-substituted phenol-formaldehyde resins were synthesized and cured by various processes, by formaldehyde or s-trioxane under acidic conditions, formaldehyde under basic conditions. The oxygen indices (OI) and char yields were determined and graphically correlated (Figure 8)(29).

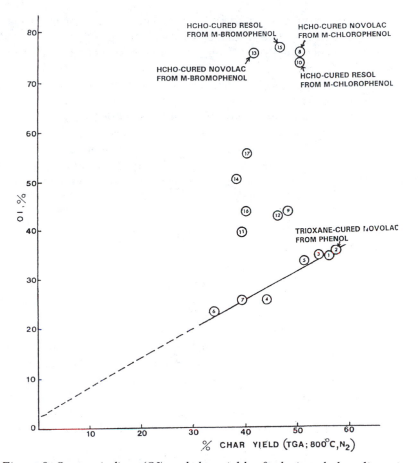

Figure 8. Oxygen indices (OI) and char yields of substituted phenolic resins. Resins in group 1-7 are halogen-free. Resins in the middle group 9,11-12,14,16-17 are based on chloro- or bromophenols but are cured with trioxane or terephthaloyl chloride. For detailed identification, see ref. 29.

Phenolic polymers are, in general, good char-formers. The particular combination of thermochemical and thermophysical properties of these char-forming polymers have made them of special interest as high-temperature-resistant, flame-retardant polymeric materials. Thermal degradation mechanistic studies of the char-forming phenol-formaldehyde-derived resins were conducted to provide information for the systematic design of high-temperature flame-resistant phenolic polymers and copolymers (*29*). The evaluation of the effect of

46

various substituents indicates unusually high oxygen indices for the halogen-substituted cured resins in relation to their char yields (Figure 8).

A set of phenolic copolymers with different weight percentage content of halogen substituted phenols was synthesized as both novolacs and resols. The results indicated no increase of oxygen index for the cured novolac copolymers, whereas an increase is observed for the cured resol copolymers. It appears that the resol process of polymerization gave products with better flame resistance (*29*). The extremely high oxygen indices of the halogen-substituted phenolics, even in relation to their char yields, indicates another effective strategy of designing "fire-smart" polymers, namely, the building-in of more than one mode of flame retardancy such as a propensity to char formation and the possibility of concurrent vapor phase action.

Comments

One practical shortcoming of the "fire-smart" polymer approach to flame retardancy is that polymer manufacturers usually prefer to optimize their products for properties other than flame retardancy. The introduction of structures specifically to favor flame retardancy may entail a compromise of other important properties. Moreover, putting in specialized functional groups often means increased cost. Therefore, we suggest a strategy of finding a means to utilize, as crosslinking sites, those latent functional groups or reactive sites that are already present in commercial polymers, such as cyano groups, ester groups, amide groups or double bonds. This strategy suggests research in the direction of finding latent catalysts and/or polyfunctional latent crosslinking reactants.

Another strategy which we consider promising is the blending of the flammable polymer with a small but effective amount of a highly flame retardant additive oligomer or polymer, which may be designed to have compatibilizing groups.

References

(1) Pearce, E.M.; Khanna, Y.P.; Raucher, D. In *Thermal Characterization of Polymeric Materials;* Turi, E.A., Ed.; Academic Press: New York, **1981**; Chapter 8, pp. 793-843.
(2) Pearce, E.M. In *Contemporary Topics in Polymer Science,* Vol. V; Vandenburg, E.J., Ed.; Plenum: New York, **1984**; pp. 401-413.
(3) Chaudhuri, A.K.; Min, B. Y.; Pearce, E. M. *J. Polym. Sci.: Polym. Chem. Ed.* **1980**, *18*, 2949-2958.

(4) Khanna, Y.P.; Pearce, E.M.; Forman, B.D.; Bini, D.A. *J. Polym. Sci.: Polym. Chem. Ed.* **1981**, *19,* 2799-2816.

(5) Khanna, Y.P.; Pearce, E.M. *J. Polym. Sci.: Polym. Chem. Ed.* **1981**, *19,* 2835-2840.

(6) Khanna, Y.P.; Pearce, E.M. *J. Appl. Polym. Sci.* **1982**, *27,* 2053-2064.

(7) Kapuscinska, M.; Pearce, E.M. *J. Polym. Sci.: Polym. Chem. Ed.* **1984**, *22,* 3989-3998.

(8) Kapuscinska, M.; Pearce, E.M.; Chung, H.F.M.; Ching, C.C.; Zhou, Q.X. *J. Polym. Sci: Polym. Chem. Ed.* **1984**, *22,* 3999-4009.

(9) Pearce, E.M. *Pure & Appl. Chem.* **1986**, *58,* 925-930.

(10) Whang, W.T.; Kapuscinska, M.; Pearce, E.M. *J. Polym. Sci.: Polym. Symp.* **1986**, *74,* 109-123.

(11) Kim, S.; Pearce, E.M. *Makromol. Chem., Suppl.* **1989**, *15,* 187-218.

(12) Kim, S.; Pearce, E. M.; Kwei, T. K., *Polym. Adv. Technol.* **1990**, 1, 49-73.

(13) Whang, W.T.; Pearce, E.M. In *Fire and Polymers*, ACS Symposium Series 425; Nelson, G.L., Ed.; ACS, Washington, D.C., **1990,** pp. 266-271.

(14) Khanna, Y. P.; Pearce, E. M. in *Flame Retardant Polymeric Materials*, Lewin, M.; Atlas, S.; Pearce, E. M., Eds., Plenum Press, New York, 1978; Vol. 2, pp 43-61.

(15) Brauman, S. K. *J. Polym. Sci.: Chem. Ed.*, **1979,** 17, 1129-1144.

(16) Clubley, B. G.; Boyce, I. D.; Hyde, T. G.; Lamb, F.; Randell, D. R. U.S. Patent 4,248,976, 1981.

(17) Oh, S.Y.; Pearce, E.M. *Polym. Adv. Technol.* **1993**, *4,* 577-582.

(18) Oh, S. Y.; Pearce, E.M.; Kwei, T.K. in *Fire and Polymers*, ACS Symposium Series 599; Nelson, G.L., Ed.; Am. Chem. Soc.: Washington, D.C., **1995,** pp. 136-158.

(19) Zeldin, M.; Jo, W. H.; Pearce, E. M. *J. Polym. Sci., Polym. Chem. Ed.* **1981**, *19,* 917-923.

(20) Yan, H.-J.; Pearce, E.M. *J. Polym. Sci.: Polym. Chem. Ed.* **1984**, *22,* 3319-3334.

(21) Li, M.Y.; Pearce, E.M.; Reiser, A.; Narang, S. *J. Polym. Sci.: Part A: Polym. Chem.* **1988**, *26,* 2517-2527.

(22) Pearce, E.M.; Lin, S.C.; Lin, M.S.; Lee, S.N. in *Thermal Methods in Polymer Analysis*, Shalaby, S.W., Ed.; Franklin Institute Press: **1978**; pp. 187-198.

(23) Chen, C.S.; Bulkin, B.J.; Pearce, E.M. *J. Appl. Polym. Sci.* **1982**, *27,* 1177-1190.

(24) Chen, C.S.; Bulkin, B.J.; Pearce, E.M. *J. Appl. Polym. Sci.* **1982**, *27,* 3289-3312.

(25) Lo, J.; Pearce, E.M. *J. Poly. Sci.: Polym. Chem. Ed.* **1984**, *22,* 1707-1715.

(26) Chen, C.S.; Pearce, E.M. *J. Appl. Polym. Sci.* **1989**, *37,* 1105-1124.

(27) Liu, Y.-I.; Pearce, E.M.; Weil, E.D. *J. Fire Sci.* **1999**, *17,* 240-258.

(28) Lu, S.; Melo, M.M.; Zhao, J.; Pearce, E.M.; Kwei, T. K. *Macromolecules* **1995**, *28*, 4908-4913.

(29) Zaks, Y.; Lo, J.; Raucher, D.; Pearce, E.M. *J. Appl. Polym. Sci.* **1982**, *27,* 913-930.

Chapter 5

High Performance Polyimide Foams

Martha K. Williams[1], Gordon L. Nelson[2], James R. Brenner[2],
Erik S. Weiser[3], and Terry L. St. Clair[3]

[1]Labs and Testbeds, National Aeronautics and Space Administration,
YA-F2-T, Room 3141, O & C Building,
Kennedy Space Center, FL 32899
[2]Florida Institute of Technology, 150 West University Boulevard,
Melbourne, FL 32901
[3]Langley Research Center, National Aeronautics and Space Administration,
6A West Taylor Street, Building 1293A,
Room 120A, Mail Stop 226, Hampton, VA 23681-0001

Aromatic polyimides have been attractive for applications in the aerospace and electronics industries. Unique properties such as thermal and thermo-oxidative stability at elevated temperatures, chemical resistance, and mechanical properties are common for this class of materials.[1] Newer to the arena of polyimides is the synthesis of polyimide foams.[2] In the present work, three different, closely related, polyimide foams are comparatively studied, including thermal, mechanical, surface, flammability, and degradation properties. In previous studies, the data relate to films[3] and not foams. Foams have much higher surface areas and are a greater challenge to fire retard. Understanding degradation and properties versus structure, foam versus solid, is of interest. Data indicate that subtle differences in chemical structure result in large differences in surface area, which further result in large differences in heat release and other flammability properties.

NASA Langley Research Center (LARC) has been developing the next generation of polyimide foam materials that will be utilized for such things as cryogenic insulation, flame retardant panels and structural sub-components.[2] This new foam technology allows for the processing of polyimide neat or syntactic foams, foam-filled honeycomb or other shapes, and microspheres, all of which produce useful articles. These products can be used in a variety of ways: flame retardant materials and for fire protection, thermal and acoustic insulation, gaskets and seals, vibration damping pads, spacers in adhesives and sealants, extenders, and flow/leveling aids. This process can also produce polyimide foams with varying properties from a large number of monomers and monomer blends. The specific densities of these foams can range from 0.008 g/cc to 0.32 g/cc.[2]

Although certain physical and mechanical properties of some of these polymers have been previously studied and reported, the characterization and understanding of such properties as flame retardancy, fire protection, liquid oxygen (LOX) compatibility, and stability versus degradation of the foams are now of significant interest. The study of their respective properties (comparative) with regards to seemingly minor changes in chemical structure, density and surface area present a novel area of research. The information gained should be of significant value in the application of polyimide foams and polymeric foams in general.

Figure 1 includes the chemical structures of the foams (see Table I for abbreviations). The last letter of each foam name designation identifies density values such as H for the higher density and L for the lower density foam.

Experimental

Material Synthesis

The salt-like foam precursor was synthesized by mixing monomer reactants of a dianhydride with a foaming agent (tetrahydrofuran) in methanol at room temperature. The dianhydride (e.g., ODPA, 756g-2.4 moles) was dispersed in a mixture of tetrahydrofuran (THF-480g) and 280 g of methanol (MeOH) at room temperature. This solution was treated at 70 °C for 6 hours in order to convert the ODPA into ODP-DEDA (dialkylester-diacid) complexed with THF by hydrogen bonding. Stoichiometric amounts of the diamine (e.g., 3,4'ODA, 488 g-2.4 mol) were added to the resulting solution of ODPA-DEDA and stirred for 2 hours to yield a homogeneous polyimide precursor solution. The resulting polyimide precursor solution had a solids content of 70 wt % and a viscosity of 20 poise at 20 °C. The solution was then charged into a stainless-steel vat and treated at 70 °C for 14 hours in order to evaporate the solvent (THF and MeOH). The resulting material was crushed into a fine powder (2 to 500μm) and sieved using a mesh, if needed. The polyimide precursor powder was then treated for an additional amount of time (0 to 300 minutes) at 80 °C to further reduce the residual solvents to around 1-10 wt% depending on the final density desired. Residual amounts of THF were determined by measuring the proton NMR spectra of the powders.[4]

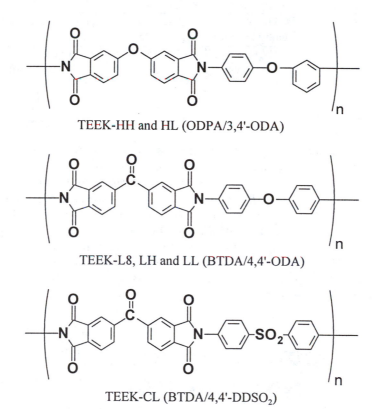

TEEK-HH and HL (ODPA/3,4'-ODA)

TEEK-L8, LH and LL (BTDA/4,4'-ODA)

TEEK-CL (BTDA/4,4'-DDSO₂)

Figure 1. Chemical structures of foams.

Table I. Foam Materials Description

Foam, Density	Description
TEEK-HH (0.082 g/cc)	ODPA/3,4'-ODA (4,4 oxydiphthalic anhydride/3,4-oxydianiline)
TEEK-HL (0.032 g/cc)	ODPA/3,4'-ODA (4,4 oxydiphthalic anhydride/3,4-oxydianiline)
TEEK-L8 (0.128 g/cc)	BTDA/4,4'-ODA (3,3,4,4-benzophenenone-tetracarboxylic dianhydride/4,4-oxydianiline)
TEEK-LH (0.082 g/cc)	BTDA/4,4'-ODA (3,3,4,4-benzophenenone-tetracarboxylic dianhydride/4,4-oxydianiline)
TEEK-LL (0.032 g/cc)	BTDA/4,4'-ODA (3,3,4,4-benzophenenone-tetracarboxylic dianhydride/4,4-oxydianiline)
TEEK-CL (0.032 g/cc)	BTDA/4,4'-DDSO$_2$ (3,3,4,4-benzophenenone-tetracarboxylic dianhydride/4,4-diaminodiphenyl sulfone)

Foam Fabrication

The amount of polyimide foam precursor powder necessary to completely foam a desired volume was placed in a chamber. The mold, shown in Figure 2, was heated to 140 °C for 60 minutes using heat plates on the top and bottom. The mold was then rapidly transferred to a nitrogen-convection oven set at 300 °C and held at 300 °C for 60 minutes. The mold was then cooled to room temperature. At this point the foam was post cured for several hours at 200 °C to remove all trace volatiles. The foam was removed from the mold and was ready for use.[4] The foams were supplied by NASA Langley Research Center and Unitika LTD, Japan. Table I lists the abbreviations for the materials and densities of the foams studied.

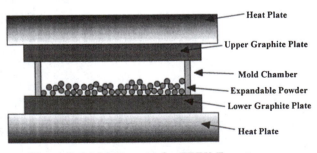

Figure 2. Mold Concept for TEEK Foaming.

Results

Mechanical Properties

Polyimide foams at varying densities were tested for mechanical and physical properties and reported.[2] Results are included in Table II. In the lower density materials, the TEEK-H and TEEK-L series have similar tensile strengths, with the TEEK-C series giving lower values comparatively. As expected, higher density increases both tensile strength and compressive strength for the same series of polymer.

Table II. Mechanical Properties

Property	TEEK-HH	TEEK-HL	TEEK-LL	TEEK-CL
Tensile Strength	1.2 MPa	0.28 MPa	0.26 MPa	0.09 MPa
Compressive Strength	0.84 MPa @10% Defl.	0.19 MPa @10% Defl.	0.30 MPa @10% Defl.	0.098 MPa @10% Defl.
Compressive Modulus	6.13 MPa	3.89 MPa	11.0 MPa	

Thermal Stability

Thermal properties of the polymers were studied in a comparative process using a TA Instruments Hi-Res Model 2950 Thermogravimetric Analyzer (TGA). TGA will measure the percentage of weight loss at different temperatures and the weight of residue at high temperature. Data reported in Table III were carried out in air, with a heating rate of 50.0 °C/min from room temperature to 800 °C with a resolution of 5 (associated with percent/minute value that will be used as the control set point for furnace heating). Analyses in nitrogen versus air indicate greater stability in nitrogen (agrees with previously reported data for polyimide films).[3] A TA Instruments Differential Scanning Calorimeter Model (DSC) 2920 was used to determine glass transition temperature, T_g (see Table III), and other thermal changes.[5] Results were obtained by heating foam samples 20.0 °C/min to 350 °C, holding at 350 °C for 10 min, equilibrating at 25.0 °C, and reramping at 20.0 °C/min to 350 °C, using the reramp curve to calculate T_g.

Data indicate that at 50% weight loss, the TEEK-CL series has the highest thermal stability, with the other series very similar in stability. It also has the highest T_g. The TEEK-L series ranks next, with the TEEK-H series having the lowest T_g of 237 °C. Previous research on polyimide films indicates that although the dianhydride structure does not have a profound impact on the polyimide stability, some consistencies or trends in the ranking of the dianhydrides have been seen, depending upon whether thermal analysis was carried out dynamically or isothermally.[6] Diamine structure appears to have a greater influence on the stability, with the ranking being $SO>SO_2>CH_2>O$. [3]

Table III. Thermal Properties for TEEK Polyimide Foams

Property	Test Method	TEEK-HH	TEEK-HL	TEEK-L8	TEEK-LH	TEEK-LL	TEEK-CL
Density G/cc	ASTM D-3574 (A)	0.08	0.032	0.128	0.08	0.032	0.032
Thermal Stability, Temp.°C	10% wt loss	518	267	522	520	516	528
	50 % wt loss	524	522	525	524	524	535
	100% wt loss	580	578	630	627	561	630
Glass Transition Tg, Temp.°C	DSC	237	237	283	278	281	321

Flammability Testing

Oxygen Index (OI), ASTM D2863, and Glow Wire Ignition, ASTM D6194

Table IV indicates the results of OI and glow wire testing at 960 °C. The OI measures the minimum amount of oxygen required for sustained combustion. Data range from 42 to 51, with higher densities slightly increasing the OI for the same series of materials (LL, LH, and L8).

In comparing different polymers, ASTM D6194, Glow-Wire Ignition of Materials, measures the ignition behavior of insulating materials using a glowing heat source. The samples did not ignite, but the depths of penetration are reported where contact with the heated source was made. The TEEK-HL indicated the most penetration when compared to the lower density materials, with the higher density TEEK-L series having the least amount of penetration.

Table IV. Oxygen Index and Glow Wire

	TEEK-HL	TEEK-HH	TEEK-LL	TEEK-LH	TEEK-L8	TEEK-CL
OI	42	49	49	51	51	46
Glow Wire Depth of penetration	most	Less than HL	Less than HH	Less than LL	Best, less than LH	Compares to HH

Radiant Panel, ASTM E162

The Radiant Panel was designed to measure both critical ignition energy and rate of heat release, the reporting value being the flame spread or radiant panel index, I_s.[7] Analyses reveal that the flame spread index, I_s, is close to zero for all of the foams evaluated. However, the percentage of shrinkage (Figure 3) and the amount of charring are different. Considering the lower density materials, 0.032g/cc, TEEK-CL reveals the least amount of shrinkage, and TEEK-HL the most. In comparing the same chemical structure, more shrinkage and degradation are observed in TEEK-HL (0.032g/cc) versus TEEK-HH (0.08g/cc), see also Figures 4 and 5. Figures 4 and 5 are photographs of duplicate 6X18 inch foam samples exposed to Radiant Panel testing.

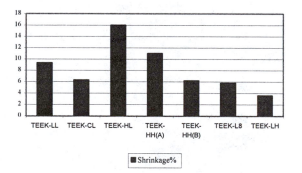

Figure 3. Shrinkage % from Radiant Panel samples.

As can be readily seen in the shrinkage data of the foams, chemical structure or density does not directly explain the differences observed. Given these results another correlation was investigated, namely surface area/porosity.

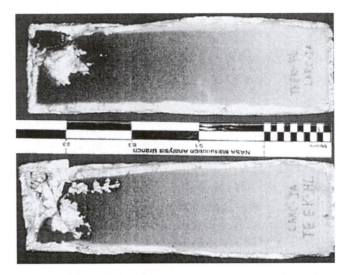

Figure 4. Duplicate samples, 6X18 inch foam TEEK-HL
(0.032 g/cc) exposed to Radiant Panel.

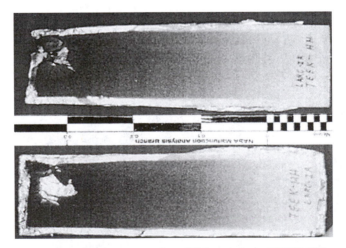

Figure 5. Duplicate samples, 6X18 inch foam TEEK-HH
(0.08 g/cc) exposed to Radiant Panel.

Surface Area/Porosity

The surface areas were measured by the single point BET method using a Quantasorb Model QS-17 sorption analyzer.[8] Foam samples were weighed and then inserted between plugs of Pyrex wool inside a 6 mm ID Pyrex U-shaped tube. The samples were pretreated by heating to 150 °C for 10 minutes using a heating mantle under flowing He (99.998%), at 200 cc/min (298 K, 1 atm) to remove any adsorbed moisture. Following pretreatment and cooling to room temperature under He, the adsorbate gas mixture of 29.3% N_2 in He was flowed over the foam. While the adsorbate mixture was flowing over the foam, the foam sample and enclosing Pyrex U-tube were submerged in liquid N_2 (77 K), and the thermal conductivity detector (TCD) signal was collected until a steady-state was achieved and integrated automatically. Duplicate experiments were performed, and the reported values are the averages. Surface area calculations and percentage of shrinkage gave a direct correlation as observed in Figure 6.

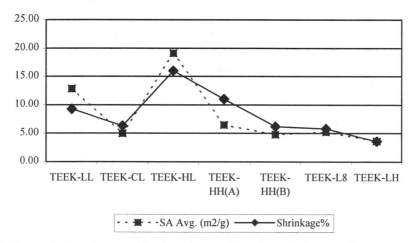

Figure 6. Correlation of Surface Area vs Shrinkage % from Radiant Panel.

Cone Calorimetry

Cone analysis was performed per ASTM E1354. Cone analysis utilizes oxygen consumption during combustion as a measure of heat release. Table V presents cone data, including time to ignition (Time ig), peak heat release rate (PHRR), average heat release rate (ave. HRR), total heat release rate (THR), smoke as specific extinction area (SEA), carbon monoxide production (CO), average mass loss rate (MLR), and initial and final masses.[9]

Table V. Cone Calorimeter Data

Sample	TIME ig (s)	PHRR (kW/m²)	Ave. HRR (kW/m²)	THR (MJ/m²)	Ave. SEA (m²/kg)	Ave. CO (kg/kg)	Ave. MLR (g/s/m²)	Initial Mass(g)	Final Mass (g)
TEEK-HL 35 kW/m²		11.5	3.10	2.3	530	0.31	0.35	9.2	7.5
TEEK-HL 50 kW/m²	89.8	54.7	21	12.9	169	0.38	1.86	13.4	0.0
TEEK-HL 75 kW/m²	13.8	154	28.9	22.5	160	0.13	1.22	14.1	4.7
TEEK-HH 35 kW/m²		7.93	3.27	2.43	618	0.07	0.32	19.8	18.8
TEEK-HH 50 kW/m²	55.0	50.5	24.8	37	261	0.19	1.18	24.6	6.7
TEEK-HH 75 kW/m²	12.6	79.7	35.1	49.1	72.9	0.02	1.39	23.7	4.3
TEEK-LL 35 kW/m²		17.3	1.43	5.25	32.9	0.00	3.80	9.4	7.2
TEEK-LL 50 kW/m²	32.6	39.9	25.2	22.7	126	0.28	1.05	13.6	4.5
TEEK-LL 75 kW/m²	8.8	53.4	26.3	21.2	117	0.03	1.12	13.6	4.7
TEEK-LH 50 kW/m²	43.8	30.7	20.3	43.0	334	0.34	0.96	25.6	4.8
TEEK-LH 75 kW/m²	17.5	43.7	25.4	53.9	6.14	0.14	1.07	26.0	3.8
TEEK-L8 50 kW/m²	37.6	36.2	26.1	54.2	2.06	0.27	1.12	29.8	6.4
TEEK-L8 75 kW/m²	15.5	58.3	30.1	68.3	33.5	0.15	1.75	34.2	4.3
TEEK-CL 35 kW/m²		5.93	1.8	1.26	226	0.00	0.25	8.1	7.4
TEEK-CL 50 kW/m²	133	26.2	17.9	15.4	55.8	0.28	0.92	12.1	4.9
TEEK-CL 75 kW/m²	9.6	68.6	27	19	149	0.03	1.28	13.1	4.0

The cone analyses indicate that there is no ignition occurring at 35 W/m², with not even complete combustion at 75 kW/m². At 50 kW/m², TEEK-CL has the longest time to ignition at 133 seconds, followed by the TEEK-H series. However, at 75 kW/m², the times to ignition are similar. Table VI summaries the PHRR data observed at the different heat fluxes for the samples analyzed.

Table VI. PHRR and Ignition (Sec) Data From Cone Analysis

Sample	Density g/cc	Surface Area	PHRR 75 kW/m²	PHRR 50 kW/m²	PHRR 35 kW/m²
TEEK-HH	0.08	5.5	80 13ig	51 55ig	8
TEEK-HL	0.032	19.1	154 14ig	55 89ig	11
TEEK-L8	0.128	5.2	58 16ig	36 38ig	
TEEK-LH	0.08	3.6	44 18ig	31 44ig	
TEEK-LL	0.032	12.9	54 9ig	40 33ig	17
TEEK-CL	0.032	5.0	69 10ig	26 133ig	6

In Figures 7 and 8, the correlation between PHRR and surface area are presented in chart form. As observed in Figure 7, except for the TEEK-LL and TEEK-CL samples, surface area gives a fairly direct correlation with the PHRR rates at 75 kW/m². These data indicate that even with a significant increase in surface area for the TEEK-L series, the PHRR rates were not entirely proportional. However, as observed in Figure 8, the data collected at 50 kW/m² were more proportional. The average heat release data (ave. HRR) are similar for all the samples, with TEEK-CL slightly lower at the 50 kW/m² heat flux. The total heat release data (THR) appear to correlate well with densities of samples. The mass loss data (MLR) are similar throughout the samples. The specific extinction area (SEA, smoke) data for the samples at 35 kW/m² are a lot higher (due to the non-flaming pyrolysis), with the higher density materials giving lower values at 75 kW/m² (flaming).

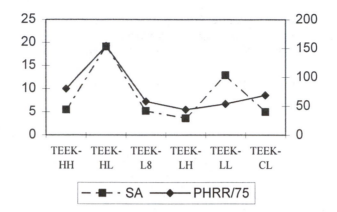

Figure 7. Comparison of PHRR at 75k W/m² and surface area.

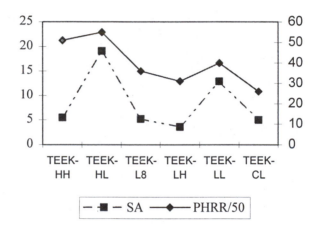

Figure 8. Comparison of PHRR at 50k W/m² and surface area.

Conclusions

Data presented validate that these newly developed polyimide foams are high performance polymers in mechanical, physical and thermal properties. Their intrinsic flame retardant nature also classifies them as fire resistant polymeric foams.

Although polyimide properties and stability versus degradation have been studied previously, the majority of the data related to films and not foams. Foams have much higher surface area and thus are a greater challenge to fire retard. Because of the intrinsic flame retardant properties of polyimides, this research has given insight into the direct correlation of chemical structure, surface area and flame retardancy of foams. While subtle changes in chemical structure undoubtedly play a role, radiant panel and cone calorimeter performance indicates that differences in the surface area or cell size of the foams appear to have a larger effect than the densities or differences in chemical structure. Chemical structure, however, may dictate surface area or porosity in the formation of foams. For example, the TEEK-CL with its SO_2 linkage, shows the lowest surface area, the lowest heat release, and highest thermal stability.

Acknowledgements

Special thanks to Langley Research Center and Unitika LTD, Japan, for providing the polyimide foams. Also thanks to NASA, Kennedy Space Center, and the National Space Club/Hugh L. Dryden Memorial Fellowship for doctoral study support.

Special thanks to colleagues J. Eggers, P. Faughnan, and R. Frankfort.

Special acknowledgement to J. Haas, S. Motto, J. Taylor and S. Bailey at White Sands Test Facility, NM, for carrying out the cone analyses.

This document was prepared under the sponsorship of the National Aeronautics and Space Administration. Neither the United States Government nor any person acting on behalf of the United States Government assumes any liability resulting from the use of the information contained in this document, or warrants that such use will be free from privately owned rights.

The citation of manufacturer's names, trademarks, or other product identification in this document does not constitute an endorsement or approval of the use of such commercial products.

References

1. Hou, T.H; Chang, A.C.; Johnson, N.J.; St.Clair, T.L. Processing and Properties of IM7/LARC-IAX2 Polyimide Composites. *Journal of Advanced Materials* **1996**, *27,* 11-18.
2. Hou, T.H.; Weiser, E.S.; Siochi, E.J.; St. Clair, T.L. and Grimsley, B.W. *Processing and Properties of Polyimide Foam,* 44th International SAMPE Symposium, 1999, 1792-1806.
3. Cella, J.A. *Degradation and Stability of Polyimides,* Marcel Dekker, Inc., New York, 1996, 343-365.

4. Weiser, E.S.; Johnson, T.F.; St. Clair, T.L.; Echigo, Y.; Kaneshiro, H.; Grimsley, B. High Temperature Polyimide Foams for Aerospace Vehicles, *Journal of High Performance Polymers,* March **2000**, Vol. 12, 1, 1-12.
5. Pramoda, K.P; Chung,T.S.; Liu, S.L.; Oikawa, H; and Yamaguchi, A. Characterization and Thermal Degradation of Polyimide and Polyamide Liquid Crystalline Polymers, *Polymer Degradation and Stability* **2000**, *67,* 365-374.
6. Wright, W.W. *Dev. Polym. Degrad.* **1981**, *3,* 1.
7. Nelson, G.L. *Fire and Polymers, Materials Tests for Hazard Prevention;* American Chemical Society: Washington, D.C., 1995; p 11-14.
8. Gregg, S. J.; Sing, K. S. *Adsorption, Surface Area and Porosity;* Academic Press, Inc., 1982; p 303.
9. Wilkie,C.A.; Grand, A.F. *Fire Retardancy of Polymeric Materials*; Marcel Dekker, Inc.: NY, 2000; p 102-107.

Chapter 6

Comprehensive Study of Thermal and Fire Behavior of *para*-Aramid and Polybenzazole Fibers

Xavier Flambard[1], Serge Bourbigot[1,*], Sophie Duquesne[2], and Franck Poutch[3]

[1]Laboratoire de Génie et Matériaux Textiles (GEMTEX), UPRES EA2161, Ecole Nationale Supérieure des Arts et Industries Textiles (ENSAIT), BP 30329, 59056 Roubaix Cedex 01, France
[2]Laboratoire de Génie des Procédés d'Interactions Fluides Réactifs-Matériaux (GEPIFREM), UPRES EA2698, Ecole Nationale Supérieure de Chimie de Lille (ENSCL), BP 108, 59652 Villeneuve d'Ascq Cedex, France
[3]Centre de Recherche et d'Etude sur les Procédés d'Ignifugation des Matériaux (CREPIM), Parc de la Porte Nord, Rue Christophe Colomb, 62700 Bruay-la-Buissière, France

In this work, we study the thermal and fire degradation of poly(p-phenylene-2,6-benzobisoxazole) (PBO) fibers in comparison with traditional p-aramide (PPT) fibers. The superior performance of PBO in comparison with PPT fibers in terms of heat resistance and flame retardancy is demonstrated. During the degradation of the fibers in a well-ventilated room, only CO, CO_2 and H_2O are observed in the gas phase. In the condensed phase, PPT and PBO fibers decompose to polyaromatic compounds which are able to trap free radicals.

Introduction

The field of high performance fibers has witnessed considerable growth in the last three decades (1). A large number of high performance polymeric fibers are on the market today and they exhibit many enhanced properties in comparison with the traditional polymeric materials, such as high mechanical properties, heat resistance or good flame retardancy. Their applications are numerous (fragmentation barrier, protective clothing, heat barrier, ...). Poly(p-phenylene-2,6-benzobisoxazole) (PBO) registered under the trademark Zylon® is a new high performance fiber (1). It is a polybenzazole containing an aromatic hetero-cyclic ring (Figure 1). It is a rigid rod isotropic crystal polymer.

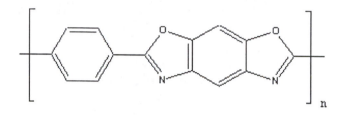

Figure 1. Repeat unit of PBO

The polybenzazoles have been developed by US Air Force researchers as super heat resistant polymer which could surpass the traditional aramide fibers. PBO has superior tensile strength and modulus compared to the classical p-aramide fibers (poly-p-phenylenediamine-terephtalamide fibers (PPT)). In our laboratory, we have recently shown that this fiber, as a knitted fabric, has excellent properties in cutting and stab resistance (3). The combination of various knitted layers gives exceptional results with PBO fibers (stab-resistance of 25 Joules for a textile structure of 3 kg/m² with an English blade).

PBO has good flame resistance and thermal stability among organic fibers and, in particular, in comparison with p-aramide fibers. As an example, the limiting oxygen index (LOI measured according to NF G 07-128) of PBO is 68 vol.-% (1) whereas that of p-aramide fibers is only 30 vol.-% (4). The LOI test is not very representative of a fire, even if it enables one to quantitatively rate the materials. The approach that is developed to evaluate fibers as knitted fabric is to use the cone calorimeter as fire model.

In this paper, we will compare and will discuss the thermal and fire behavior of PBO fibers in comparison with PPT. The gas phase will be studied during the fire degradation using FTIR and the composition of the materials in the condensed phase will be investigated using solid state ^{13}C NMR.

Experimental

PBO fibers were supplied by Toyobo (Japan) and is registered under the trademark Zylon®. PPT fibers are classical Kevlar®. The yarns used in this study have the following characteristics:

PBO :   Ne 20/2 (Nm 2/34 – 60 Tex) spun yarn

PPT :   Ne 16/2 (Nm 2/28 – 70 Tex) spun yarn

PBO and PPT fibers have been knitted on an automatic flat machine gauge E7. The texture used is a double woven rib. The two samples have a surface weight equaling 1.08 kg/m² (4 yarns knitted together in the case of PPT and 5 yarns in the case of PBO).

TG analyses were performed using a Setaram MTB 10-8 thermobalance at 10 °C / min from 20°C to 1200°C under air flow (Air Liquide grade, 5×10^{-7} m³/s measured in standard conditions) and under nitrogen flow (N45 Air Liquide grade, 5×10^{-7} m³/s measured in standard conditions) . Samples (about 10 mg) were placed in vitreous silica pans; precision on temperature measurements is ±1.5°C.

The Stanton Redcroft Cone Calorimeter was used to carry out measurements on samples following the procedure defined in ASTM 1354-90. The standard procedure used involves exposing specimens measuring 100 mm x 100 mm x 2.5 mm thick in the horizontal orientation. External heat fluxes of 50, 75 and 100 kW/m² have been used; these fluxes have been chosen because 50 kW/m² is common heat flux in mild fire scenario and 75 & 100 kW/m² represents postflashover conditions (5). When measured, RHR (rate of heat release) and VSP (volume of smoke production (6)) values are reproducible to within ±10%. The results presented in the following are averages; the cone data reported in this paper is the average of three replicated experiments. The weight loss of the samples during combustion was not recorded, because of the low mass of the textile. The curves were too noisy to be used and yields of CO and CO_2 and the heats of combustion could not be computed.

Sample of decomposition gases were then taken from an exhaust pipe of the cone calorimeter continuously pumped with a capacity of 8 liters/minutes and analyzed by FTIR. A smoke filter removes soot particles from the gas sample before it reaches the FTIR analyzer. The sampling line, approximately 4 meter long, is maintained at about 183°C in order to avoid condensation. The FTIR spectrophotometer (Nicolet 710C) was placed on-line and it has enabled the continuous detection of gas components. The gas cell had a volume of approximately 750 cm³, a 3.77 meter path length and was operated at 650 torr. The FTIR was set to generate one spectrum every 10 sec.

^{13}C NMR measurements were performed on a Brucker ASX100 at 25.2 MHz (2.35 T) with magic angle spinning (MAS), high power ^{1}H decoupling (DD) and ^{1}H-^{13}C cross polarization (CP) using a 7 mm probe. The Hartmann-Hahn matching condition was obtained by adjusting the power on ^{1}H channel for a maximum ^{13}C FID signal of adamantane. All spectra were acquired with contact times of 1 ms. A repetition time of 10 s was used for all samples. The reference used was tetramethylsilane and the spinning speed was 5000 Hz. The values of chemical shift are verified using adamantane and adamantanone before starting a new experiment. For these products, the chemical shifts were within ± 0.2 ppm.

Results and Discussion

TG curves (Figure 2) show the high heat resistance of PBO fibers in comparison with PPT fibers. Whatever the atmosphere, the heat resistance of PBO fibers is always higher. Under air, PPT fibers begin to degrade at about 450°C and form a 3 wt.-% residue at 1200°C whereas PBO fibers degrade at about 600°C and form a 3 wt.-% residue at 1200°C. Under nitrogen, the degradation of fibers begins at higher temperature (about 550°C for PPT and about 700°C for PBO). Larger amounts of residue are observed at 1200°C (38 wt.-% for PPT and 65 wt.-% for PBO). This shows the strong influence of oxygen on the thermal degradation of the fibers. This last consideration is important when it is recalled that the burning process depend on the thermal degradation reactions in the condensed phase, which generate volatile products and, that the polymeric substrate heated by an external source is pyrolyzed with the generation of combustible fuel in a zone where there is depletion of oxygen (7).

Rate of Heat Release (RHR) curves (Figure 3) of knitted PPT and PBO fibers under three external heat fluxes (50, 75 and 100 kW/m²) show that PBO fibers show very good fire behavior in comparison with PPT fibers. RHR peaks of PBO under 50 kW/m², 75 kW/m² and 100 kW/m² are respectively only 60 kW/m², 150 kW/m² and 250 kW/m² respectively compared with 400 kW/m², 430 kW/m² and 650 kW/m² for PPT fibers. Moreover, the time to ignition of PBO is longer than that of PPT (i.e, 27 s vs. 54 s at 75 kW/m²). This demonstrates the high fire resistance of PBO.

Smoke obscuration is significantly lowered for PBO in comparison with p-aramide fibers (Figure 4). PPT fibers evolve smoke with a peak at 0.012 m³/s and this peak does not depend on the external heat flux, except for the time when it occurs (72s at 50 kW/m², 42s at 75 kW/m² and 25s at 100 kW/m²). PBO fibers do not contribute to the smoke obscuration during fire whereas the smoke production of burning p-aramide material is comparatively high. If fibers are to be used as flexible fire barriers, this becomes important in term of safety of people because the obscuration of a room or a corridor generally leads to panic; indeed panic gives rise to more deaths than the fire itself (8).

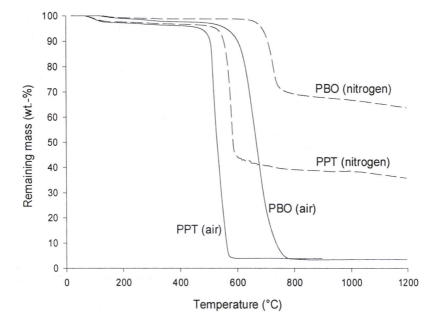

Figure 2. TG curves of PPT and PBO fibers under nitrogen (dashed lines) and air (plain lines) (heating rate = 10°C/min)

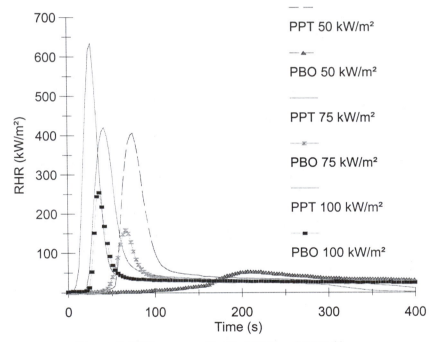

Figure 3. RHR curves of knitted PPT and PBO fibers

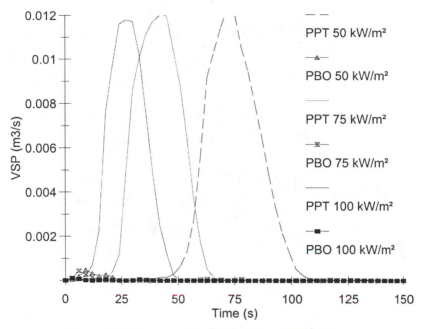

Figure 4. VSP curves of knitted PPT and PBO fibers

The evolved gases collected from the cone experiments (external heat flux = 75 kW/m²) were analyzed by FTIR (Figure 5 and Figure 6). In these conditions (well ventilated room), only CO (2000-2250 cm^{-1}), CO_2 (2250-2400 cm^{-1} and 635-720 cm^{-1}) and H_2O (3500-4000 cm^{-1} and 1350-1850 cm-^{1}) are observed. The amount of the gases is always lower in the case of the combustion of PBO (typical example in Figure 7). As the cone combustion is highly overventilated, it is not a surprise that only major products of combustion are detected. This also means that the hazard of PBO during a fire is lower than that of PPT especially, if it is used as a fire blocker.

The RHR curves of PPT and PBO fibers enable the determination of several characteristic times at an external heat flux of 75 kW/m² (Figure 3). The first event to notice is the heating of the fibers (t=20s for PPT and 30s for PBO). The ignition of the materials occurs at 27s for PPT and 54s for PBO. After the ignition, a sharp peak is observed corresponding to the combustion of the fibers (t = 33s for PPT and t = 90s for PBO). It is noteworthy that between the ignition and the end of the RHR peak, the two behaviors are very different: PPT burns with a large flame whereas PBO burns with a small flame. After the RHR peak, glowing of the fibers occurs which leads to their degradation (t ≥ 75s for PPT and t ≥ 90s for PBO).

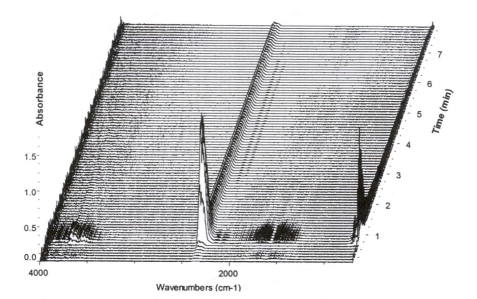

Figure 5. Infrared spectra of evolved gaseous products of PPT fibers versus time (conditions of the cone calorimeter, heat flux = 75 kW/m²)

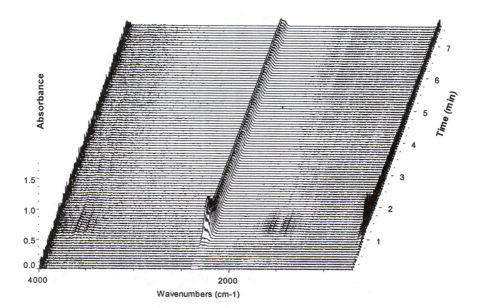

Figure 6. Infrared spectra of evolved gaseous products of PBO fibers versus time (conditions of the cone calorimeter, heat flux = 75 kW/m²)

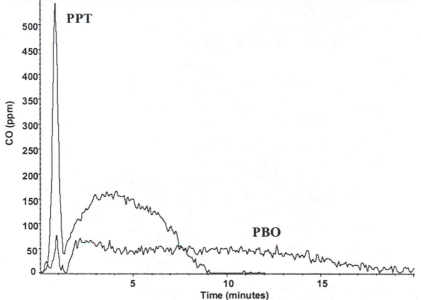

Figure 7. Evolved CO of PBO and PPT fibers versus time measured using FTIR connected to the cone calorimeter (heat flux = 75 kW/m²)

In order to understand what species are formed as a function of time during combustion in the condensed phase, the external heat flux (75 kW/m²) is shut down at the different characteristic times defined above. Then the sample is removed from the cone calorimeter and quenched in the air before ^{13}C NMR analyses. The assignments of the bands of the two fibers before degradation are shown in Figure 8 and Figure 9.

The spectra (Figure 10 and Figure 11) reveal that the fibers decompose after the RHR peaks when glowing phenomenon occurs (t ≥ 75s PPT and t ≥ 90s PBO). Indeed for the two fibers, the bands in the aromatic region (100-160 ppm) broaden and only one band is then observed. It is centered about 130 ppm and it can be assigned to the formation of aromatic/polyaromatic species[9],[10]. The width of the band suggests the presence of several non-magnetically equivalent carbons [11]. As discussed before in previous work[12], the band can be assigned to several types of partially oxidized aromatic and polyaromatic species. At long times, the loss of the signal can be assigned to the paramagnetic behavior of the sample. The fibers are transformed into polyaromatic compounds which can trap free radicals [12].

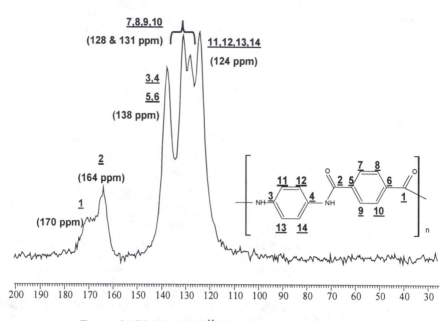

Figure 8. CP-DD-MAS ^{13}C NMR spectra of PPT fibers

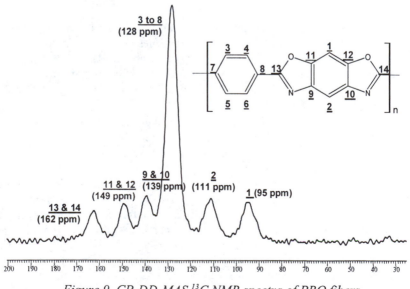

Figure 9. CP-DD-MAS ^{13}C NMR spectra of PBO fibers

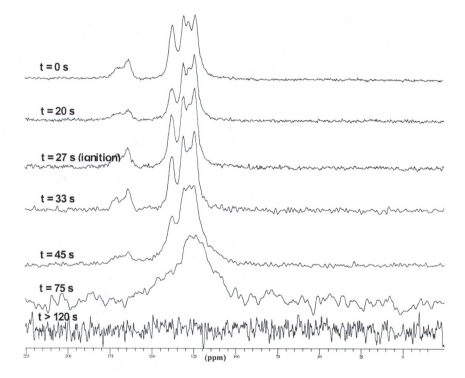

*Figure 10. ¹³C NMR spectra of PPT fibers versus characteristic time (conditions
of the cone calorimeter, heat flux = 75 kW/m²)*

74

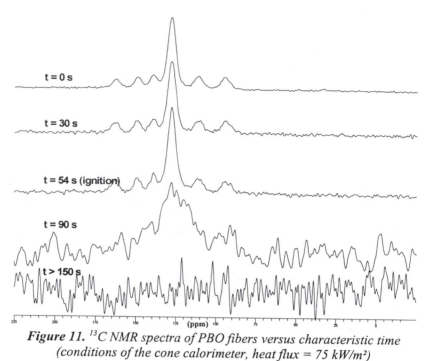

Figure 11. *^{13}C NMR spectra of PBO fibers versus characteristic time (conditions of the cone calorimeter, heat flux = 75 kW/m²)*

Conclusion

This work has demonstrated the superior performance of PBO in comparison with traditional p-aramide (PPT) fibers in terms of thermal and fire behavior. During the degradation of the fibers in the conditions of a well ventilated room, only CO, CO_2 and H_2O are observed in the gas phase. In the condensed phase, PPT and PBO fibers decompose to polyaromatic compounds which are able to trap free radicals.

Acknowledgment

The authors are indebted to Mister Dubusse and Mister Noyon from CREPIM for their skilful experimental assistance in cone calorimeter experiments. NMR experiments were made in the common research center of the University of Lille and Mister Bertrand Revel is acknowledged for helpful discussion and experimental assistance.

References

(1) Hongu, T.; Philips, G.O. *In "New Fibers"*, Woodhead Publishing Limited, Cambridge; England, **1997**.

(2) Kitagawa, T.; Murase, H.; Yabuki, K. *J. Polym. Sci. Part B : Polym. Phys.*, **1998**, *36*, 39-48.

(3) Flambard, X.; Polo, J., *In Fiber Society Spring Conference "Sustainability and recycling of textile materials"*, Ferreira, F. N. Ed., Fiber Society; Guimares, Portugal, **2000,** 147-148.

(4) Yang, H.H., *In "Kevlar aramid fiber"*, Yang, H.H. Ed., John Wiley & Sons; Chichester, **1993**, 191-192.

(5) Babrauskas, V. *Fire and Mat.*, **1984**, 8, 81-95.

(6) Babrauskas, V.; Grayson, S.J. *In "Heat Release in Fires"*, Elsevier Science Publishers Ltd; London, **1992**.

(7) Lewin, M. *In Fire Retardancy of Polymers : The Use of Intumescence*, Le Bras, M.; Camino, G.; Bourbigot, S.; Delobel, R., Eds., Royal Chem. Soc.; Cambridge, **1998**, 3-32.

(8) Akalin M.; Horrocks A.R.; Price D. *J. Fire Sci.*, **1988**, 6, 333-334.

(9) Earl, W.L.; Vanderhart, D.L. *J. Magn. Reson.*, **1982**, 48, 35-54.

(10) Supaluknari, S.; Burgar, I.; Larkins, F.P. *Org. Geochim.*, **1990**, 15, 509-519.

(11) Maciel, G.E.; Bartuska, V.J.; Miknis, F.P. *Fuel*, **1979**, 58, 391-394.

(12) Bourbigot, S.; Le Bras, M.; Delobel, R.; Descressain, R.; Amoureux, J.P. *J. Chem. Soc. - Faraday Trans.*, **1996**, *92*, 149-158.

Polyurethanes

Polyurethanes constitute a major, diverse class of plastics. In the United States and Canada alone some 6 billion pounds of polyurethanes were produced in 2000. Foam products constitute 68% of the polyurethane demand, with coatings, adhesives, sealants, and elastomers constituting only 32% collectively. End-uses for polyurethanes are very diverse, from construction and transportation, to furniture, appliances, carpet cushion, packaging, and fibers. In the United States some 2.5 billion pounds of isocyanates are used each year: 53% polymeric MDI, 9% pure MDI, 34% TDI and 4% for all specialty isocyanates. Some 58% of polymeric MDI is used in rigid foam and 92% of TDI is used in flexible foam (1,2).

In the absence of flame retardants, polyurethanes tend to be quite flammable and produce extensive smoke while burning. Numerous approaches have been used to flame retard polyurethanes, from halogen and phosphorus additives to more exotic materials like silicones and ferrocene. The first paper (Najafi-Mohajeri et al.) discusses the fire performance of representative MDI-based polyurethane elastomers and foams. A wide variety of flame retardants were used. A traditional additive like halogen/antimony oxide appeared to be more effective in elastomers while halogen/phosphorus additives seemed more effective in flexible foams. Different test protocols gave different results. While some additives gave significant reductions in peak rates of heat release as measured by the cone calorimeter, the same additives did not provide resistance to Bunsen burner flame exposure. Yet open flame exposure is probably closer to regulatory needs. Some additives in foams resulted in an increase in foam density as well as in a loss of physical properties.

Representative of efforts to find new non-halogen flame retardants for polyurethanes, the second paper (Camino et al.) uses expandable graphite. The oxygen index increases from 22 to 42 at 25 wt.% loading. Expandable graphite expands ten-fold in the same temperature range as the polymer gives major combustible products. While there is an increase in volatiles in the presence of graphite, the worm-like structure developed by the graphite suffocates the flame. Complete breakdown of polyurethane bonds is prevented by crosslinking.

Graphite helps stabilize the final char residue. The third paper (Choi et al.) discusses a novel phosphorus flame retardant for polyurethane coatings.

References

1. Lichtenberg, F. W. *Polyurethane End Use Markets, U.S. and Canada, 1998 and Projected 2000;* Alliance for the Polymers Industry, http://www.polyurethane.org/statistics_slideshow/statspips113presfinal_files/frame.htm.
2. Lichtenberg, F. W. *Polyurethanes Market Statistics;* Alliance for the Polymers Industry, http://www.polyurethane.org/statistics_slideshow/sld001.htm.

Chapter 7

Cone Calorimetric Analysis of Modified Polyurethane Elastomers and Foams with Flame-Retardant Additives

N. Najafi-Mohajeri[1], C. Jayakody[2], and G. L. Nelson[1]

[1]College of Science and Liberal Arts, Florida Institute of Technology, 150 West University Boulevard, Melbourne, FL 32901
[2]Chestnut Ridge Foam, Inc., Route 981 North, Latrobe, PA 15650

The effects of commercially available flame-retardant additives on cone calorimetric results of molded flexible polyurethane foam materials were studied. Zinc stearate at 10% by weight loading had the best results in prolonging the time to ignition (TTI). The combination of HP-36 and Sb_2O_3 had the highest percentage peak heat release rate (PHRR) reduction (44%) compared to all other additives. The presence of halogens and phosphorus compounds, that are very effective in the gas phase, caused a considerable increase in smoke and carbon monoxide production. In addition, these studies showed that the flame retardants had a much greater impact on reduction of PHRRs of elastomer materials compared to polyurethane foams.

Flexible and rigid polyurethane foams are used in a wide range of commercial applications such as seat cushions, furniture, carpet underlayment, mattress padding, and insulation material for construction and appliances. In recent years, the polyurethane foam industry has been challenged by two main issues: chlorofluorocarbon (CFC) blowing agent replacement and flammability.

The ozone depletion potential with the use of CFC blowing agents is no longer a major concern since the industry has replaced CFCs with environmentally friendly agents such as water and carbon dioxide. Presently, more than 75% of foam volume is blown with water and carbon dioxide.

The flammability of polyurethane foams and the need for new materials with improved fire retardancy have been the subjects of numerous studies. The fact that such foams burn when exposed to an ignition source and may burn quite rapidly is known to be due to the combination of high surface area, ready access of oxygen because of open cell structure and low heat capacity or thermal inertia.

It is generally accepted that Cone Calorimeter (ASTM E1354, ISO 5660) is a meaningful method to estimate hazard from a burning material. Information such as ignitability, heat release rate, smoke release, and effective heat of combustion are obtained from standard test runs using this apparatus.

Over the past several years, we have been studying new and traditional approaches to flame retardancy in polyurethane foam and elastomeric materials (*1-7*).

This study discusses and compares the effect of commercially available flame-retardant additives on cone calorimetric results for two different related systems: solid diphenylmethane-4,4'-diisocyante (MDI) based PUR elastomers and molded PUR flexible foam materials based on polymeric MDI.

Experimental

Materials

The base materials were Bayer's Bayfit® 566 A/B and Baytec® ME-090 for the water-blown polyurethane foam and elastomeric materials respectively (Table I). The Bayfit® 566B contains a mixture of polyether polyols, water as blowing agent, catalysts, surfactants, and cross-linkers. The Bayfit® 566A is a polymeric diphenylmethane-4,4'-diisocyanate (MDI) with NCO content of 32.5%. Baytec® ME-090 is a MDI/ polytetramethylene ether glycol (PTMEG) based prepolymer with NCO content of 9.93%.

The additives used in this study and their suppliers are given in Table I. Some of these additives such as Fyrol® FR2 have been specifically recommended for polyurethanes by manufacturers.

Table I. Materials and Suppliers

Chemical	Description	Supplier
Bayfit® 566A	Polymeric MDI	Bayer
Bayfit® 566B	Polyether polyol system	
Baytec® ME-090	MDI/PTMEG- based prepolymer (NCO% = 9.93)	
Dow Corning® 1-9641	Si additive/amine functional	Dow
Dow Corning® 4-7105	Si additive/non-functional	Corning
Dechlorane Plus® 25	Aliphatic chlorine-containing crystalline organic compound	Laurel Industries
Antimony Oxide	Sb_2O_3, ultrafine II Grade	
DE-83R	Decabromodiphenyl oxide (83.3% Br)	Great Lakes Chemical
DE-60F	Pentabromodiphenyl oxide blend	
Firemaster® HP-36	Halogenated phosphate ester (44.5% halogen, 7.5% P)	
Antiblaze® 100	Chloroalkyldiphosphate ester	Albright &
Antiblaze® 230	Halogenated phosphorus ester	Wilson
Fyrol® FR2	Tri(1,3-dichloro isopropyl) phosphate	Akzo Nobel Chemicals
Fyrolflex® RDP	Resorcinol bis(diphenyl phosphate)	
Synpro® Mg stearate		Ferro
Synpro® Zinc stearate		

Synthesis

All the polymer formulations were prepared using a one-shot method.

Foams: The Bayfit® 566 Component B (polyol) was stirred for three minutes using a high-speed mechanical stirrer. The Bayfit® 566 Component A was added and stirred for another 5 seconds. The mixing ratio of polyol/isocyanate was 100:48 (isocyanate index of 100%). The mixture was then poured into a preheated (120° F) aluminum mold (16" × 4" × ¾") which was treated with mold release agent, Chemtrend® MR 515. The foam was removed from the mold after 4-5 minutes and crushed to remove trapped carbon dioxide. In the case of flame-retardant foams, the additives were first mixed with the polyol(s) for 3 minutes before the addition of Component A.

Elastomers: The standard polyurethane elastomer material was prepared by adding the corresponding amount of the chain extender butanediol to ME-090 prepolymer such that the isocyanate index was 1.05. The prepolymer was placed in an 8 oz. bottle and heated at 100-110 °C for about 30 minutes. Butanediol

82

was added and stirred for about 50 seconds. The viscous mixture was then poured into a 4" x 4" preheated mold with 1/8" thickness. The specimen was cured at 100 °C for 30 minutes followed by annealing at 100 °C for another 30 minutes. Modification with additives was performed by mixing the additive with prepolymer for 30 minutes at 100 °C using the mechanical stirrer followed by mixing the corresponding amount of butanediol for 50 seconds.

Characterization

The cone calorimetric results of 4"×4"×2" foam and 4"×4"×1/8" elastomer specimens were obtained using an Atlas cone calorimeter at 25 kW/m² heat flux in a horizontal orientation. Sample size was chosen to give similar initial mass for foams and elastomers.

Results and Discussion

Table II provides a summary of the cone calorimeter data including time to ignition (TTI), peak heat release rate (PHRR), peak time for PHRR (t), average heat released (Ave. HRR), total heat released (THR), average specific extinction area (Ave. SEA), average carbon monoxide yield (Ave. CO), percentage residue and initial mass of the polyurethane foams. The additives are classified into three main categories: non-halogen, halogen, and halogen-phosphorus additives.

The first parameter considered is TTI. As shown, sample 4 was able to extend its ignition time by almost 24 times. Although zinc stearate in sample 4 had the best results in prolonging the TTI, other additives such as Mg Stearate and HP-36 were also able to show an increase of 150% and 82% respectively in TTI.

The PHRR data shows that in general modified formulations gave lower PHRR, except for samples 5 (Mg stearate), 12 (Antiblaze® 100), 15 (Fyrolflex® RDP), and 16 (Fyrolflex® RDP/HP-36). Among the non-halogen fire retardant additives, an amine functionalized siloxane additive 1-9641 (sample 2), at 3.3% wt loading, reduced the PHRR by 40%.

Some halogen and halogen-phosphorus additives were the most effective class of flame retardant for reducing PHRR. Particularly interesting appears to be the combination of HP-36 and Sb_2O_3 that showed the highest percentage PHRR reduction (44%) compared to all other additives.

Table II. Cone Calorimeter Data for Modified Foam Polyurethanes (25 kW/m²)

	Sample (% wt)	TTI (s)	PHRR (kW/m²)	t (s)	Ave. HRR (kW/m²)	THR (MJ/m²)	Ave. SEA (m²/kg)	Ave. CO (kg/kg)	Residue (%)	Initial Mass (g)
1	Base Polymer	15.6	412	176	225	57.4	413	0.02	6.6	24.6
Non-Halogen Additives										
2	1-9641 (3.3)	13.7	249	76	126	54.2	404	0.01	13.0	24.6
3	4-7105 (5)	16.0	340	131	193	66.7	459	0.02	10.4	28.9
4	Zn Stearate (10)	372.3	340	476	174	64.4	597	0.01	9.9	30.9
5	Mg Stearate (10)	39.1	444	211	194	70.8	431	0.02	11.7	31.5
6	Sb₂O₃ (20)	11.4	372	146	269	71.2	511	0.03	27.8	42.8
Halogen Additives										
7	Dec. Plus® / Sb₂O₃(15,5)	18.5	316	191	190	52.4	709	0.11	15.6	35.5
8	Dec. plus® / Sb₂O₃ / 1-9641(9,3,3)	17.6	275	66	166	53.3	739	0.07	16.4	30.7
9	DE-83R / Sb₂O₃ (12,8)	25.7	288	176	168	47.9	769	0.17	20.9	40.3
10	DE-60F (20)	19.1	259	196	118	39.5	832	0.13	8.8	28.7
Halogen-Phosphorus Additives										
11	HP-36 (20)	28.4	315	211	144	49.9	881	0.11	10.2	33.9
12	Antiblaze® 100 (20)	12.4	439	171	223	55.8	838	0.08	13.1	34.6
13	Fyrol® FR2 (20)	18.4	326	201	163	48.2	745	0.08	13.2	31.5
14	HP36/ Sb₂O₃ (28,7)	21.8	234	271	143	61.3	793	0.13	22.5	48.5
15	Fyrolflex® RDP (10)	22.6	429	151	210	56.7	793	0.08	8.2	30.1
16	Fyrolflex® RDP / HP-36 (16,4)	15.8	465	161	195	51.6	954	0.13	10.1	31.7
17	Antiblaze® 230 (9)	26.1	274	176	154	49.2	727	0.10	9.9	26.7

84

Although materials with a single step degradation mechanism without charring typically produce a single PHRR, fire retardant materials often do not produce such simple HRR-time curves. The appearance of multiple HRR peaks, the location of the maximum on the time axis and its relative magnitude to other peaks in the same plots are important features (*8*). Figure 1 shows plots of four specimens that produced different HRR-time curves.

The PHRR for sample 14 was recorded at 234 kW/m² after 271 seconds. However, the sample shows that a similar peak with almost the same magnitude had appeared only after ~90 seconds. Hence, for this sample, Ave. HRR may be a better judgment of fire performance.

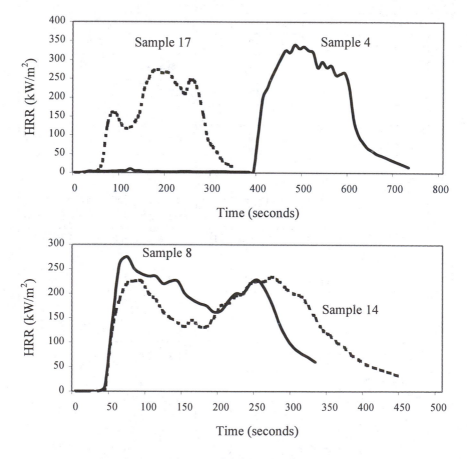

Figure 1. Cone calorimeter HRR-time curves

Sample 17 shows multiple HRR peaks compared to sample 4 that has only one peak. Sample 4 is polymer and zinc stearate.

As noted before, the location of the PHRR on the time axis is important. Although samples 8 and 17 have the same PHRR value, the positions of these peaks on the time axis were very different; PHRR for sample 8 happened after 66 seconds and for sample 17 after 176 seconds.

In a real fire situation, the most suitable material presents the lowest emission value at the longest time. Using this criterion, the ratios between the PHRR and relative peak time are shown in Figure 2. Considering that the lower the value of this ratio, the better the fire performance, samples 4 and 14 exhibited promising results. Sample 2 with the lowest PHRR within the non-halogen additive category showed the highest PHRR/t.

Concerning the THR, reported in Table II, the modified polyurethane foams compared to an unmodified sample did not show any considerable improvement except for brominated additive DE-60F that showed a 31% reduction.

Finally, the presence of halogens and phosphorus compounds, that are very effective in the gas phase, caused a considerable increase in smoke and carbon monoxide production.

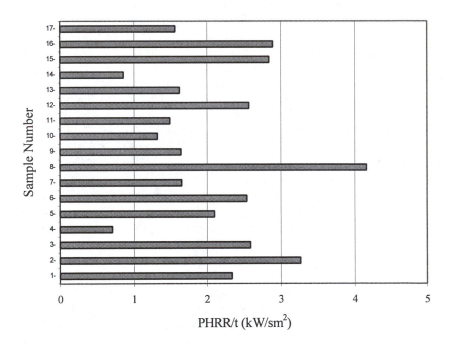

Figure 2. Ratio of PHRR to the relative peak time

Foams Versus Elastomers

The data in Table III show that the peak rate, the average rate and total amount of heat released (PHRR, Ave. HRR, and THR) during the combustion of unmodified PUR foams are lower compared to polyurethane elastomers. This phenomenon may be due to the differences in their molecular structures and starting materials. Elastomers are linear polyurethanes based on MDI / PTMEG prepolymer with free NCO content of 9.93% and 1,4-butanediol as chain extender. The PHRR of virgin PTMEG (MW 2000) is 1301 kW/m^2, which can contribute to the high PHRR values of elastomeric materials.

PUR foams are based on polymeric MDI with free NCO content of 32.5% and a blend of polyethers. The formation of amine functionalities by the reaction of excess NCO and water and the presence of cross-linkers in the Bayfit® 566 B polyol lead to more cross-links and urea linkages in foams compared to elastomers.

As shown in the Table III, all modified formulations gave lower PHRR for both foam and elastomeric materials. The reductions for PUR foams were in the range of 3% to 40% relative to the control. In contrast, all modifications of PUR elastomers with flame-retardant additives showed a larger impact on PHRR reduction compared to modified PUR foam materials. These reductions were in the range of 67% to 90%. However, cone residues were 2.5 to 5 times higher in foams than in elastomers. Higher cone residues in foams are attributed to the highly cross-linked structure of these materials.

Among the non-halogen fire retardant additives studied, an amine-functionalized siloxane additive 1-9641 (5% wt) reduced the PHRR by 79% for the polyurethane elastomer. A 3.3% wt loading of this additive in PUR foams dropped the PHRR by 40%. The same result was repeated for a non-functionalized siloxane additive 4-7105. This additive at 5% wt loading showed a 70% reduction in PHRR for the elastomer compared to only 17% reduction in PHRR for foam at the same loading level.

Halogenated additives were the most effective class of flame-retardant for reducing PHRR, Ave. HRR, and THR in PUR elastomers, but less so in foams. However, they showed the highest increase in smoke production for both foams and elastomers.

Dechlorane Plus® 25 and antimony oxide at 15 and 5% wt loading respectively, reduced the PHRR by 90% (370 vs. 3696 kW/m^2) in PUR elastomer. However, the PHRR was reduced by only 23% for foam at the same loading level. In addition, a combination of siloxane additive 1-9641 (5% wt), Sb$_2$O$_3$ (5% wt) and Dechlorane Plus® 25 (15% wt) did not show an additional decrease in PHRR in elastomers. The same additives (3%, 3%, 9% respectively) in a foam formulation showed an additional decrease to 275 kW/m^2.

Table III. Cone Calorimeter Data for Modified Foam and Elastomer Polyurethanes (25 kW/m²)

Sample (% wt)	PHRR (kW/m²)	Ave. HRR (kW/m²)	THR (MJ/m²)	Ave. SEA (m²/kg)	Ave. CO (kg/kg)	Residue (%)	Initial Mass (g)
Base Polymer Foam	412	225	57.4	413	0.02	6.6	24.6
Base Polymer Elastomer	3696	375	120	381	0.03	3.3	33.9
Non-Halogen Additives							
1-9641 (3.3) Foam	249	126	54.2	404	0.01	13.0	24.6
1-9641 (5) Elastomer	787	185	89.1	495	0.02	6.7	33.2
4-7105 (5) Foam	340	193	66.7	459	0.02	10.4	28.9
4-7105 (5) Elastomer	1118	212	72.3	548	0.02	5.6	27.3
Halogen Additives							
Dec. Plus®/Sb$_2$O$_3$ (15,5) Foam	316	190	52.4	709	0.11	15.6	35.5
Dec. Plus®/Sb$_2$O$_3$ (15,5) Elastomer	370	173	52.6	807	0.12	8.4	32.7
Dec. plus® / Sb$_2$O$_3$ / 1-9641 (9,3,3) Foam	275	166	53.3	739	0.07	16.4	30.7
Dec Plus® / Sb$_2$O$_3$ / 1-9641 (15,5,5) Elastomer	370	184	70.1	770	0.11	12.1	34.9
DE-83R / Sb$_2$O$_3$ (12,8) Foam	288	168	47.9	769	0.17	20.9	40.3
DE-83R / Sb$_2$O$_3$ (11,4) Elastomer	723	183	56.4	964	0.17	8.3	37.6
Halogen-Phosphorus Additives							
HP-36 (20) Foam	315	144	49.9	881	0.11	10.2	33.9
HP-36 (15) Elastomer	1039	260	79.9	664	0.13	3.6	39.8
Fyrol® FR2 (20) Foam	326	163	48.2	745	0.08	13.2	31.5
Fyrol® FR2 (14) Elastomer	1034	322	90.4	618	0.09	5.6	41.9

Among halogen-phosphorus additives, Fyrol® FR2 was one of the other additives that showed a higher percentage PHRR reduction in PUR elastomers than in foams. Fyrol® FR2 at 14% wt loading reduced the PHRR by 72% for the solid elastomer, but at 20% wt loading the PHRR reduction was only 21% for foam material. Similar behavior was shown with HP-36.

If one looks at average HRR rather than PHRR one sees little difference. The lowest average HRR for foams was given with the silicone/silica additive 1-9641. The lowest average HRR for solid elastomers was given with Dechlorane Plus/Sb_2O_3.

Dechlorane Plus®/Sb_2O_3 showed the lowest value for Total Heat Release (THR) for solid elastomers. For foams bromine or halogen/phosphorus additives gave lowest THR; Dechlorane Plus®/Sb_2O_3 gave only marginal improvement in THR. Silicone/silica additive, 4-7105, showed an increase in THR for foam material.

Smoke was increased by the presence of halogen as was carbon monoxide. Halogen-phosphorus additives were comparable to halogen in smoke production for foam formulations but were somewhat lower in elastomers. All additives showed some increase in cone residue, but that increase was not as dramatic as the difference between elastomers and foam.

Conclusions

Cone calorimetric studies of commercially available flame-retardant additives in PUR foam were conducted. Zinc stearate in 10% wt loading showed interesting fire performance by prolonging the time to ignition and reducing the PHRR by 18%. Among the non-halogen fire retardant additives, 1-9641 additive reduced the PHRR by 40% compared to an unmodified sample.

Samples 4 and 14 exhibited the best PHRR/t results and therefore they have the lowest emission value at the possible longest time.

As expected the presence of halogens and phosphorus compounds caused a significant increase in smoke and carbon monoxide production.

These studies showed that flame retardants had a much greater impact on reduction of peak heat release rates of elastomer materials compared to polyurethane foams. On average, the PHRR of modified PUR elastomers were impacted 3 times more than the PHRR of modified PUR foams. Despite the use of a number of different additives, it is clearly difficult to get PHRR values below 250 kW/m^2 and THR value below 50 MJ/m^2 for MDI-based polyurethanes.

Acknowledgment

The authors wish to thank Mr. Scott Baton, Chairman/President, and Mr. Dennis R. Robitaille, President retired, of Chestnut Ridge Foam, Inc., Latrobe, PA; and Mr. Usman Sorathia at the Office of Naval Research, West Bethesda, MD, for sponsoring this research program on flame-retardant polyurethanes. Thanks are also due to Mr. Stephen Lewandowski and Mr. Jay M. Cappelli of Bayer Corporation, Pittsburgh, PA, for their cooperation.

References

1. Benrashid, R.; Nelson, G.L.; Linn, J.H.; Wade, W.R. *J. Appl. Polym. Sci.* **1993**, *49*, 523-527.
2. Benrashid, R.; Nelson, G.L. *J. Appl. Polym. Sci. Part A: Polym. Chem.* **1994**, *32*, 1847-1865.
3. Nelson, G.L.; Jayakody, C.; Sorathia, U. *The 8th Annual BCC Conference on Flame Retardancy: Recent Advances in Flame Retardancy of Polymeric Materials,* Stamford, CT, **1998**.
4. Sorathia, U.; Jayakody, C.; Nelson, G.L. *A Cone Calorimetric Study of Flame Retardant Elastomeric Polyurethanes Modified With Siloxanes and Commercial Flame Retardant Additives,* NCWCCD-TR-64-98/06, Naval Surface Warfare Center, West Bethesda, MD, **1998**.
5. Jayakody, C.; Nelson, G.L.; Sorathia, U; Lewandowski, S. *J. Fire Sci.* **1998**, *16*, 351-382.
6. Nelson, G.L.; Najafi-Mohajeri, N.; Jayakody, C. *Fire Retardant Chemicals Association,* Proceeding, Tucson, AZ, October 24-27, **1999**.
7. Nelson, G.L.; Jayakody, C.; Myers, D. *29th International Conference in Fire Safety,* San Francisco, CA, January 10-14, **2000**.
8. Checchin, M.; Cecchini, C.; Cellarosi, B.; Sam, F.O. *Polym. Deg. Stab.* **1999**, *64*, 573.-576.

Chapter 8

Mechanism of Expandable Graphite Fire Retardant Action in Polyurethanes

G. Camino[1], S. Duquesne[2], R. Delobel[2], B. Eling[3], C. Lindsay[3], and T. Roels[3]

[1]Dip Chimica IFM, Università Di Torino, Via P. Giuria, 7, 10125 Torino, Italy
[2]GéPIFREM, ENSCL, BP 108, 59652 Villeuneuve d'Ascq, France
[3]ICI Polyurethanes, Everslaan 65, 3078 Everberg, Belgium

This study deals with the effects of expandable graphite (EG) on the mechanism of degradation of polyurethane (PU). We have shown that the mechanism of expansion of the graphite is due to oxidation of the carbon layer by H_2SO_4, rather than its decomposition, as previously suggested in the literature. On the other hand, we have shown that the mechanism of degradation of PU is little affected by the presence of EG. However, H_2SO_4 induces additional reactions which do not modify the final structure of the residue. Finally, the expansion of the PU/EG formulation generates, on the surface, an insulative layer which protects the underlying material, resulting in the fire retardant properties of interest.

Polyurethanes (PU) are different from most plastic in that they represent a large family: from foam insulation to shoe soles, car seals to abrasion-resistant coatings. Rigid foam is one of the most effective practical insulation materials used in applications such as building, pipes or domestic refrigerators. The use of PU in building is restricted because, in case of fire, it acts as a fuel generator with subsequent propagation of the fire to the surrounding combustibles and finally the destruction of property and the loss of human lives by direct burns or inhalation of toxic smoke.

In order to reduce the flammability of PU, flame retardants are added to the polymer. The most widely used FR additives in rigid foam systems are chlorinated phosphate esters. However, in case of fire, halogenated fire retarded polymers evolve more toxic combustion products than the untreated polymer (*2-3*). Moreover, generation of highly corrosive hydrogen halides occurs. Therefore, legislation tends to limit the use of halogenated fire retardant systems for environmental and safety reasons (*2-4*). These are the reasons why current research focuses on the development of halogen-free, non-toxic and environmentally friendly fire retardants and in particular on intumescent systems (*5,6*). Intumescence describes a material which forms a blown cellular charred layer upon heating. The creation of this layer limits the heat and mass transfer to and from the underlying material, thus insulating the substrate from the heat source. Intumescent fire retardant systems have been developed using expandable graphite (EG) which is capable of imparting fire retardancy to various materials when incorporated into them (*7-11*). In particular, it provides flame retardancy of interest in flexible polyurethane foams (*12-15*), acting to smother burning as well as to insulate the foam from the heat source. The purpose of this paper is to investigate the effect of EG on the degradation mechanism of PU coating. This coating may be applied on various materials, such as polymers or composites and, in particular, on rigid PU foam.

Experimental

Materials

Raw materials were commercial polymeric diisocyanate diphenylmethane (PMDI) and polyester polyol from ICI. The PU coating is obtained by polycondensation of the isocyanate with the polyol without further addition of additives or catalyst apart from those present in the proprietary commercial products. The molar ratio NCO/OH has been fixed at two. The components were stirred (400 rpm) in disposable paper cups (500 ml) at room temperature for 1 min and allowed to polymerize for 24 h.

Expandable graphite (Callotek 500, Graphitwerk Krophfmuhl) is a graphite intercalation compound (GIC). The first known phenomenon of intercalation explains the secret of the production of the fine Chinese porcelain, seven centuries before Christ (*16*). However, the first GIC was only prepared in 1841, accidentally, by Schafhautl while analyzing crystal flakes of graphite in a solution of sulfuric acid (17). The graphite structure consists of layers of hexagonal carbon structures within which a chemical compound (e.g. H_2SO_4) can be intercalated (Figure 1). EG is prepared either by oxidation with a chemical reagent (*18*) or electrochemically (19-20) in the intercalating acid (i.e. H_2SO_4, HNO_3, etc). The chemical reaction in the case of H_2SO_4 is expressed by the following equation:

$$24nC + mH_2SO_4 + 1/2 O_2 \longrightarrow C^+_{24n}(HSO^-_4)(m\text{-}1)H_2SO_4 + 1/2H_2O \quad (1)$$

The result is a graphite acid salt, so named because of its ionic nature in which the positive charge of the oxidized graphite network is balanced by negatively charged acid anions (*211,222*) and also includes acid molecules (*233*) as shown in Figure 1 for H_2SO_4 GIC. Elemental analysis shows that Callotek 500 contains 2.8 wt.-% S corresponding to 8.5 wt.-% H_2SO_4.

The PU/EG (15 wt.-%) coating was made in the same way as the PU coating (see above), using polyol containing EG prepared by stirring both components 1 min at 4000 rpm using a Turbo Turrax apparatus.

Thermogravimetry (TG)

Thermogravimetric analysis has been carried out either at a heating rate of 10°C/min or in isothermal conditions under nitrogen (60 cm^3/min) using a horizontal thermobalance Du Pont MOD.951 Thermogravimetric Analyzer. Samples (about 10 mg) were held in open silica pans.

Pyrolysis Unit

Pyrolysis of the materials (samples of about 100-200 mg) in an open glass holder has been carried out in the pyrolysis unit shown in Figure 2, using nitrogen as a carrier gas, either in isothermal or programmed heating conditions. Aluminum foil is wrapped around the walls of a water cooled trap inside the degradation tube in order to collect the high boiling products (HBP), which are volatile at reaction temperature but nonvolatile at ambient temperature. Gaseous degradation products are trapped from the carrier gas using the gas trapping system of Figure 3 which allows direct IR analysis through the KBr windows or sampling for GC-MS through the septum.

Fourier Transformed Infra-red spectroscopy

IR spectra of original and degradation products were recorded using a Perkin-Elmer FTIR 2000 spectrometer connected to a Grams Analyst 2000 Perkin Elmer data station. Solids were ground and mixed with KBr to form pellets. Liquids and gases were respectively examined between KBr discs or in the gas cell of Figure 3.

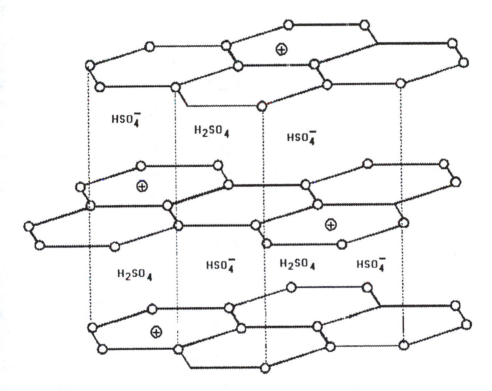

Figure 1: Structure of H₂SO₄ graphite acid salt.

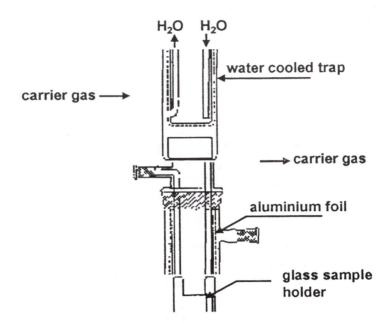

H_2O H_2O

water cooled trap

carrier gas →

→ carrier gas

aluminium foil

glass sample holder

Figure 2: Glass pyrolysis unit.

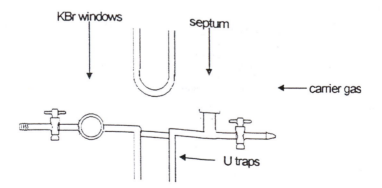

KBr windows   septum

← carrier gas

← U traps

Figure 3: Gas trapping system.

Combustion behavior

Minimum oxygen concentration (OI) for self-sustained combustion of a vertical down burning specimen after top ignition was measured by the OI method (ASTM D2863/77) using a Stanton Redcroft apparatus on specimens 100 x 10 x 3 mm^3.

Blowing Measurement

Blowing was measured as a function of temperature at a heating rate of 20°C/min using a homemade apparatus previously described (244). Briefly, a pellet of the material (diameter: 15 mm, thickness: 2mm) on which a position transducer probe (diameter: 20 mm) is located is heated in a glass tube (diameter: 22mm). The probe displacements are recorded as a function of temperature.

Results and Discussion

Decomposition of EG

The thermogravimetry of the EG (Figure 4) had to be carried out step by step in the pyrolysis unit raising the temperature 10°C/min and weighing the residue. Indeed, the thermobalance could not be used because of the EG expansion (about ten times) blowing the sample away from the holding pan.
The major weight loss of EG that occurs between 200°C and 350°C with a maximum rate at about 250°C is responsible of the sample expansion, which begins at 200°C. Half and total expansion is respectively reached at 260°C and 350°C. In the literature (255), it is suggested that expansion occurs via the sulfuric acid decomposition, according to:

$$H_2SO_4 \longrightarrow SO_3 + H_2O$$
$$SO_3 \longleftrightarrow SO_2 + 1/2 \; O_2 \quad (2)$$

However, sulfuric acid decomposition temperature (about 340°C) (26) is higher than that of which expansion begins. Moreover, the original amount of H_2SO_4 in the EG is on the order of 8.5 wt.-%, whereas the weight loss after the expansion in Figure 4 is about 25%; this cannot be explained by absorbed water.
In order to better understand the expansion mechanism, we have collected the gases evolved under isothermal conditions (T=250°C, 30 min) and analyzed them using IR spectroscopy. The gas phase of EG pyrolysis is composed of SO_2 (1400-1300 cm^{-1}), CO_2 (720-635, 2400-2250 and 3700-3600 cm^{-1}) and water (1850-1350 and 4000-3600 cm^{-1}). These gases escape through the edges of the

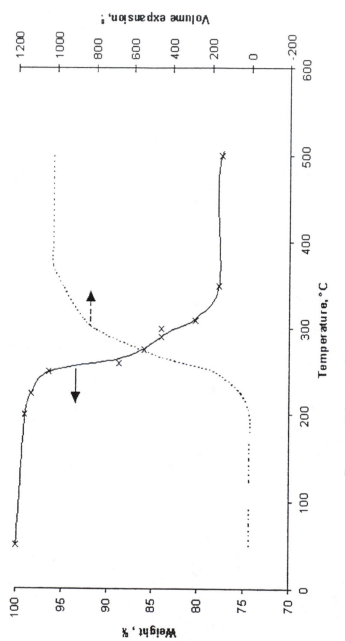

Figure 4: TG (–) and blowing measurement (- -) of EG.

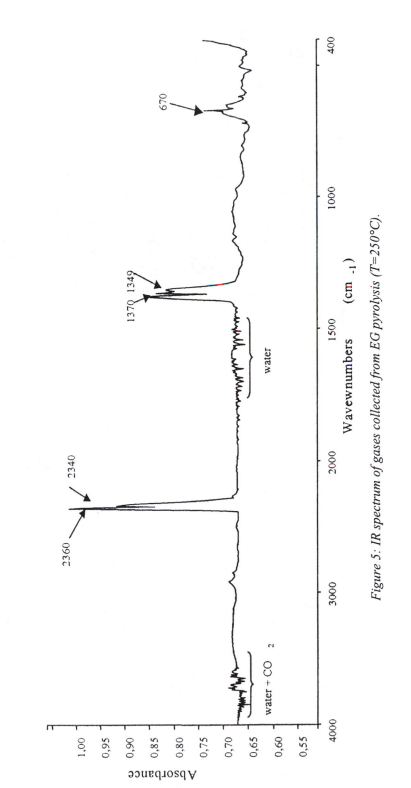

Figure 5: IR spectrum of gases collected from EG pyrolysis (T=250°C).

graphite particles, leading to the irreversible expansion.

Rather than H_2SO_4 decomposition, these data clearly show that a redox process between H_2SO_4 and graphite (reaction 3) generates the blowing gases.

$$C + 2H_2SO_4 \longrightarrow CO_2 + 2H_2O + 2SO_2 \quad (3)$$

A similar redox process was previously reported in the literature for SO_3 intercalated graphite as the process promoting exfoliation upon heating (277) :

$$3C + 4SO_3 \rightarrow S_{(S)} + 3CO_2 + 3SO_2 \quad (4)$$

Partial oxidation of the C to CO could be ruled out by a thermal volatilization analysis experiment (TVA (288)) in which non-condensable CO would have been easily detected. The quantity of SO_2 evolved was evaluated by bubbling the carrier gas into aqueous 1% H_2O_2 solution, resulting in the formation of H_2SO_4 which was then titrated with NaOH. The result was confirmed by elemental analysis of original and residual EG which, however, does not rule out occurrence of reaction (3) because we did not perform analysis for elemental sulfur independently from H_2SO_4 determination. EG evolves about 0.8 wt.-% of SO_2 corresponding to only about 16 % of the weight of the H_2SO_4 intercalated in the graphite which participates in the expansion process as shown in reaction (3). This is qualitatively confirmed by the presence of residual sulfate IR bands (1160-1060 cm^{-1} (29)) in the EG residue after the isothermal treatment, which is comparable to that of original EG. In these spectra, the IR absorptions of water (3410 and 1650 cm^{-1}) are also visible. These data confirm that only a minor fraction of the intercalated H_2SO_4 takes part in the expansion mechanism by the redox reaction (3). Further volatilization of H_2SO_4 at higher temperature ($>340°C$) does not apparently contribute to expansion (see Figure 4) possibly because the gases evolved in the redox process (200-$350°C$) have pushed the crystalline graphite sheets apart freeing the intercalated H_2SO_4. The material generated by heating EG to expansion temperature is a puffed-up material of low density with a worm like structure.

Degradation of polyurethane coating

The weight loss of the PU coating is shown to take place in three steps, as observed from the TG curve of Figure 6.

This curve shows that the PU coating decomposes to volatile products with the main step between 220 and 450°C. The residue from the first step is relatively stable and slowly further volatilizes at higher temperatures with a maximum rate at 530°C (2nd step). Above 610°C (3rd step), weight loss occurs at a steady rate and is still not complete at 800°C.

In order to characterize the products of the degradation, we have repeated every degradation step to completion under isothermal conditions (Table 1). For each step, gases, HBP and residue have been examined using IR spectroscopy.

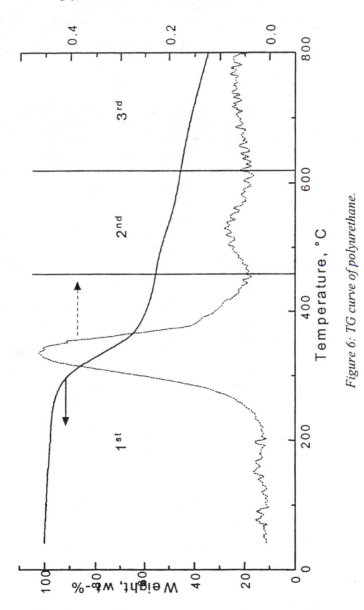

Figure 6: TG curve of polyurethane.

Table 1: PU coating thermal volatilization.

Step of degradation	Thermogravimetry			Isothermal degradation in pyrolysis unit	
	ΔT, °C	Peak, °C	Weight loss, %	Temperature, °C	Time, min
1	220-450	334	44,5	330	40
2	450-610	530	9,5	440	60
3	610-800	/	11	550	70

Comparison of the IR spectrum of the original PU with that of the residue after the first degradation step (Figure 7) shows that the urethane bands at 3330, 1720, 1070 cm^{-1} decrease significantly. The strong band around 2270 cm^{-1}, which corresponds to residual isocyanate, disappears and new bands appear at 2120-2100 and 1640 cm^{-1}, respectively, attributed to carbodiimide and urea. The bands attributed to ester and ether bonds (respectively around 1250 and 1110 cm^{-1}) are still preserved. Consequently, these functionalities seem to be slightly affected by the degradation in this step of the pyrolysis. Carbon dioxide is the dominant gas evolved during the first degradation step. The IR spectrum of the HBP shows absorption bands similar to those of the original product, apart from free isocyanate which is absent. Therefore, it is mainly fragments of polymer chain which constitute this fraction of the degradation products.

These results show that the first PU degradation step consists of the reverse of the reaction of polycondensation, leading to the formation of alcohol and isocyanate groups in agreement with the literature data (*300*). The isocyanate resulting from depolymerization and the unreacted original isocyanate, which are very reactive at degradation temperature, dimerize with evolution of carbodiimide and CO_2. A reaction between carbodiimide and water has been reported (*300*); a similar reaction of carbodiimide (reaction (5)) with alcohol may be assumed. This reaction leads to the formation of substituted urea.

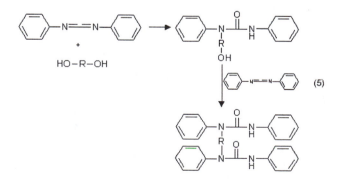

Since polyols freed by depolymerization are bifunctional, they lead to crosslinking, and consequently to the stabilization of the material. This may explain why the degradation slows after the first step (T>400°C).

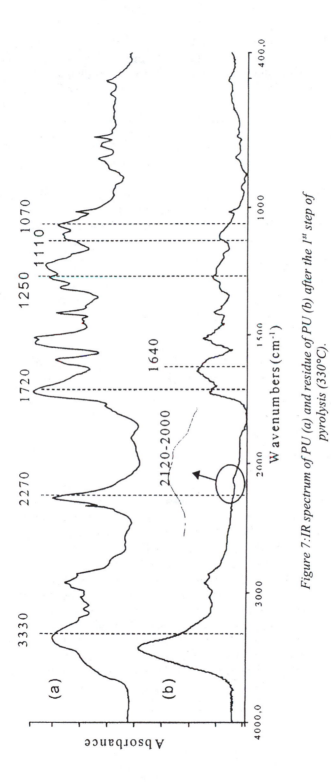

Figure 7: IR spectrum of PU (a) and residue of PU (b) after the 1st step of pyrolysis (330°C).

A reaction of water with isocyanate to give aromatic amines and carbon dioxide, or with carbodiimide to give substituted ureas might be envisaged as previously described in the literature (*300*). However, by using a water sensitive probe, we could detect water evolution only above 400°C. This agrees with the fact that dehydration of alcohols without catalyst usually occurs at a temperature around 400°C. We can not exclude the trimerization of the isocyanate groups, either original or obtained from the depolymerization, to give crosslinked thermally stable isocyanurate rings, which show typical IR absorption bands around 1700, 1410 and 730 cm^{-1}, overlapping the absorptions of PU. All these reactions are summarized in Figure 8.

During the second degradation step (T=440°C), a small weight loss takes place (less than 10 wt.-%) and the material degrades slowly. This degradation step (maximum at 530°C) should correspond to the total decomposition of the carbonyl-containing structure apart from urea bonds (i.e. residual urethanes, esters...) as shown by disappearance of the IR absorption at 1720 cm^{-1} in the residue of the second degradation step. The IR analysis of the HBP evolved in this step indicates the presence of an aromatic amine type structure, because the spectrum presents bands around 1640 cm^{-1} (N-H stretching) and 1514 cm^{-1} (aromatic structure). CO_2, which may be evolved from ester decomposition, and H_2O, which may derive from the dehydration of alcohol groups, are the gaseous components.

In the last degradation step (T=550°C), when the material degrades slowly, the urea bonds decompose leading to the formation of a charred carbonaceous structure in which urea groups have disappeared, possibly by decomposition to nitriles (IR absorption at 2225 cm^{-1}).

Degradation of PU/EG-15 wt.-%

Comparison of the thermogravimetric curves of PU and PU/EG (Figure 9) shows that EG somewhat modifies the degradation of the PU. The weight loss begins at about the same temperature in both cases (around 200°C). However, the TG of the mixture PU/EG shows an additional step between 380°C and 470°C. This leads to a decrease of the residual mass at 470°C, although the final weight loss at 800°C is the same for PU and PU/EG.

The first degradation step is similar to that of PU as shown by the similar weight loss in Figure 9 and IR spectra of the residue and the HBP. An additional absorption in the IR spectrum of the gases was found around 1350 cm^{-1} and was assigned to SO_2, as expected by the decomposition of EG discussed above. The IR of the gases collected in the second degradation step, which corresponds to the additional step compared to PU, shows the presence of CO_2 and SO_2. Moreover, an additional peak around 1736 cm^{-1} is observed. It can be attributed to compounds containing carbonyl groups, which can be evolved from the oxidation of the polymer chain by H_2SO_4.

Table 2: PU/EG coating thermal volatilization

Step of degradation	Thermogravimetry			Isothermal degradation in pyrolysis unit	
	$\wedge$T, °C	Peak, °C	Weight loss, %	Temperature, °C	Time, min
1	200-380	340	36,2	300	60
2	380-470	390	12,8	380	30
3	470-650	550	10,5	425	80
4	650-800	/	1,4	600	30

An interaction between the EG and the thermal degradation of the PU is shown by the difference in the HBP spectra. In particular, an absorption is seen in the HBP spectrum collected during the second step of PU/EG pyrolysis at 2660 cm^{-1} corresponding to an SH bond derived from a reaction between the PU and H_2SO_4 of EG which is reduced to a thiol group. Furthermore, substituted quinoline and aniline have been detected in the high boiling products using GC-MS. In the literature (*300*), the formation of aniline derivatives was previously attributed to acid catalysis (reaction 6) during thermal degradation of polyurethanes in the presence of phosphoric acid.

The spectrum of the residue is similar to that obtained for PU. Consequently, the presence of EG seems to induce additional degradation of the polymer leading to a decrease of the residual weight at 470°C. However, the mechanism of crosslinking in the condensed phase leading to the stabilization of the material seems not to be affected by the presence of EG. In the third step of the degradation, products are similar to those obtained from PU. However, the IR spectrum of the gases still shows a band around 1736 cm^{-1}, which might be attributed to carbonyl groups. Finally, the last step of degradation corresponds to a very small weight loss (around 1%). Consequently, this step of degradation occurs very slowly without major modifications of the degradation products as compared to the corresponding step for PU.

Combustion performance of PU/EG

When PU/EG is exposed to heat, expansion of the structure occurs, but in lower proportion than for pure EG. The resulting material is composed of "worms" of carbon embedded in a degraded matrix of polymer. This high expanded layer gives fire retardant properties to PU as shown by progressive oxygen index which increases from 22 to 42 vol.-% at 25 wt.-% loading (Table 3).

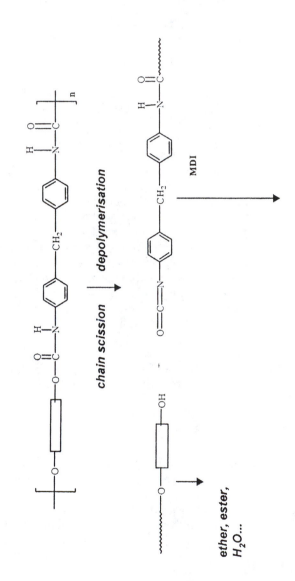

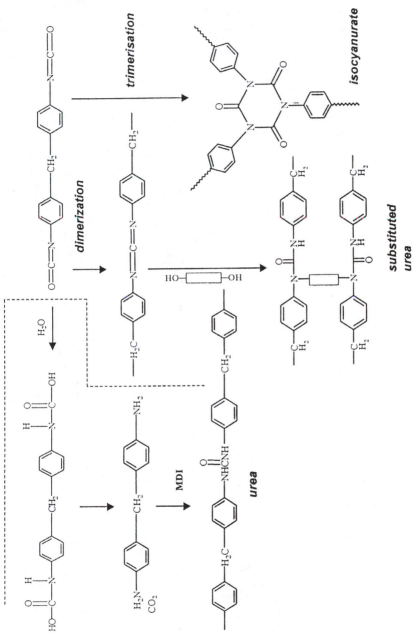

Figure 8: Mechanism of degradation of PU.

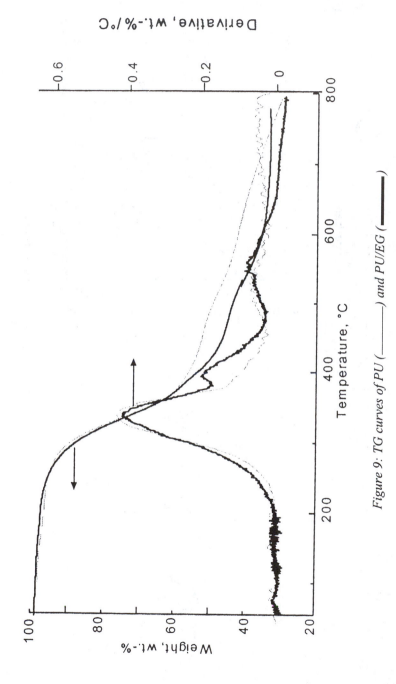

Figure 9: TG curves of PU (———) and PU/EG (———)

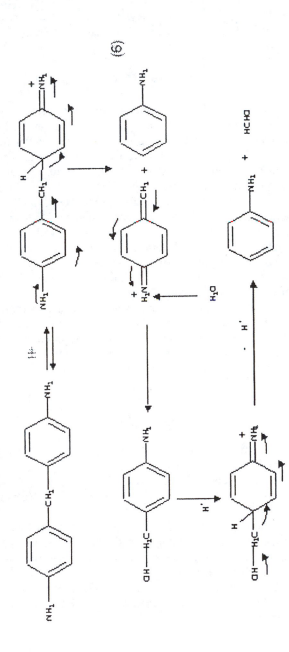

Table 3: LOI of PU/EG versus additive loading

EG loading (wt.-%)	0	10	15	20	25
OI (vol.-%)	22	26	30	38	42

Conclusion

It is shown that the volume of sulfuric acid intercalated in expandable graphite increases the volume of the material by ten times on heating above 200°C. The thermal expansion is due to gases evolved in the redox reaction between H_2SO_4 and the graphite leading to CO_2 and SO_2 with about 25 wt.-% loss. A polyurethane coating was effectively fire retarded by addition of EG which expands in temperature range of 200-300°C in which the major decomposition of the polymer to combustible products occurs. The voluminous worm-like structure developed by graphite expansion suffocates the flame. This physical effect overcomes the increase of volatiles evolved in the presence of graphite in the major step of PU decomposition. Indeed, complete break down of the PU to volatile fragments above the temperature of urethane bond depolymerization to isocyanate and alcohol is prevented by crosslinking. Dimerization of isocyanates to carbodiimide followed by addition to alcohol and the formation of substituted urea groups, can explain the thermal stabilization through crosslinking. In the presence of graphite, a further limited weight loss takes place due to oxidation and fragmentation of the polymer by the H_2SO_4 in EG. However, the resulting charred residue is more stable than in the absence of graphite and its amount at 800°C is comparable.

References

1. Woods, G. *The ICI Polyurethanes Book – 2nd Ed.*, ICI Polyurethanes and John Wiley and Sons Pub., **1990**.
2. Lomakim, S.M.; Zaikov, G.E. *New Concepts in Polymer Science*, Ecological Aspects of Polymer Flame Retardancy, VSP Utrecht, The Nederlands, **1999**.
3. Camino, G.; Luda, M.P.; Costa, L. *Chemical Industry and Environment*, Vol. I, General Aspects – Risk Analysis, Casal, J., Ed. **1993**, pp 221-227.
4. DGXI Proposal for a directive on waste from electrical and electronic equipment, Second Draft, **1999**.
5. Vandersall, H.L. *J. Fire and Flammability*, **1971**, *2*, 97-140.
6. Camino, G. *Plastics Additives*, Pritchard, G. Ed., Chapman and Hall, London, **1998**, pp 297-306.

7. Penszek, P.; Ostrysz, R.; Krassowski, D. *Flame Retardants 2000*, London, England, 8-9th Feb. **1999**, pp 105-111.
8. Krassowski, D.W.; Hutchings, D.A.; Quershi, S.P. *Fire Retardant Chemicals Association, Fall Meeting*, Naples, Florida, **1996**, pp 137-146.
9. Krassowski, D.W.; Ford, B.M. *Fire and Materials Conference*, San Antonio, Texas, **1997**.
10. Okisaki, F. *Fire Retardant Chemicals Association, Spring Meeting*, San Francisco, California, **1997**, pp 11-24.
11. Goto, M.; Tanaka, Y.; Koyama, K. European patent EP 0 824 134 A1, **1997**.
12. William, R.; Bell, H. US patent 2 168 706 A, **25/06/1986**.
13. Heitmann, U. European Patent EP 0 450 403 A3, **1992**.
14. Bell, R.W.H. GB patent 2 168 706A, **1986**.
15. Pollock, M.W.; Wetula, J.J.; Ford, B.M. US Patent 5,443,894, **22/08/1995**.
16. Weiss, A. *Angewandte Chemie*, **1963**, *2*, 697-748.
17. Schafhautl, P. *J. Für Praktische Chemie*, **1840**, *21*, 129-157.
18. Kang, F.; Zhang, T.Y.; Leng, Y. *J. Phys. Chem. Solids*, **1996**, *6-8*, 889-892.
19. Kang, F.; Leng, Y.; Zhang, T.Y. *J. Phys. Chem. Solids*, **1996**, *6-8*, 883-888.
20. Kang, F.; Zhang, T.Y.; Leng, Y. *Carbon*, **1997**, *35*, 1167-1173.
21. Zabel, H.; Solin, S.A. *Graphite Intercalation Compounds I*, Springer-Verlag, Berlin, **1990**, pp 1-3.
22. Ebert, L.B. *Annu. Rev. Mater. Sci.*, **1976**, *6*, 182-211.
23. Selig, H.; Ebert, L.B. *Advan Inorg Radiochem*, **1980**, *23*, 289-290.
24. Bertelli, G.; Camino, G.; Marchetti, E.; Costa, L.; Casorati, E.; Locatelli, R. *Poly. Deg. Stab.*, **1989**, *25*, 277-292.
25. Herold, A.; Petitjean, D.; Furdin, G.; Klatt, M. *Materials Science Forum*, **1994**, *152-153*, 281-288.
26. The Merck Index, 12th ed., Merck Research Laboratories Division of MERCK & CO., Inc Pub., **1996**, pp 1535.
27. Furdin, G. *Fuel*, **1998**, *77*, 479-485.
28. McNeill, I.C. *Developments in Polymer Degradation −1*, N. Grassie Ed., Applied Science Pub., **1977**, pp 43-66.
29. Bellamy, L.J. *The Infra-red Spectra of Complex Molecules, 2nd Ed.*, Methuen & Co Ltd, London, **1958**, pp 345.
30. Grassie, N.; Zulfiqar, M. *J. Polym Sci.: Polym Chem Ed*, **1978**, *16*, 1563-1574.

Chapter 9

Synthesis of Phosphorus-Containing Two-Component Polyurethane Coatings and Evaluation of Their Flame Retardancy

Yong-Ho Choi[1], Dae-Won Kim[1], Wan-Bin Im[2], Jong-Pyo Wu[1], and Hong-Soo Park[1]

[1]Division of Ceramic and Chemical Engineering, Myongji University, Yongin 449-728, Korea
[2]National Institute of Technology and Quality, Kwacheon 427-010, Korea

Pyrophosphoric lactone modified polyester containing two phosphorus functional groups in one structural unit of base resin was synthesized to prepare a non-toxic reactive flame-retardant coating. The pyrophosphoric lactone modified polyester was cured at room temperature with isocyanate and isophorone diisocyanate (IPDI)-isocyanurate to get a two-component polyurethane flame-retardant coating (TAPPU). Three kinds of flammability tests were conducted, 45° Meckel burner, limiting oxygen index (LOI), and Cone calorimetry. These results indicate that the coatings offer good flame retardancy; the flame retardancy of these coatings was increased with the content of phosphorus.

Until now many of the flame-retardant coatings were halogen-containing compounds, but, because of the environmental pollution problem, regulations to limit the amount of toxic gas emission, such as halogenated materials, will be enforced. One of the most promising non-toxic coatings will probably be one that contains phosphorus. It is known that flame-retardant coatings containing phosphorus show 2~4 times more flame-retarding effect than those containing bromine or chlorine functional groups(1). Ma et al.(2) prepared a polyurethane flame-retardant coating with spirocyclic phosphate polyol, and they found that flame retardancy was increased with the content of spirocyclic phosphate polyol. Liu et al.(3) prepared polyurethane coatings containing phosphorus by reaction of bis (4-isocyanatophenoxy) phenyl phosphine oxide with polyol. Weil and McSwigan(4) prepared flame-retardant coatings by blending of phosphoric melamine salt with other resins. Most of the flame-retardant coatings are prepared not by a reaction but by blending so that cracks on the film surface due to a phase separation and blooming phenomenon are observed.(5) Though a few reactive flame-retardant coatings containing phosphorus were developed as mentioned above, they contain one phosphorus functional group in their structure unit so that it is difficult to get an efficient flame-retarding effect.

In this study to obtain a non-toxic reactive flame-retardant coating containing phosphorus moieties in the structural unit, we synthesized pyrophosphoric lactone modified polyester that contains two phosphorus functional groups in its base resin structure unit. Then, the pyrophosphoric lactone modified polyester was cured at room temperature with isocyanate, isophorone diisocyanate (IPDI)-isocyanurate, to get the two-component polyurethane flame-retardant coating. The physical properties and flame retardancy of the coatings are examined to validate the possibility of their use as flame-retardant coatings.

Experimental

Reagents

The reagents used in this study, pyrophosphoric acid (PYPA; Aldrich Chemical Co.), adipic acid (AA; Sigma Chemical Co.), trimethylolpropane (TMP; Tokyo Kasei Kogyo Co.), and 1,4-butanediol (BD; Junsei Chemical Co.) were all reagent grade, and polycaprolactone 0201[PCP; MW 530, OH No. 212, viscosity (at 55°C) 65 cP, Union Carbide Co.] was purified grade. Desmodur Z-4470 [Z-4470; IPDI-isocyanurate type, solid content 70.5%, NCO content 11.8%, viscosity (at 23°C) 1600±700 mPa·S, Bayer Leverkusen Co.] was used as isocyanate; TiO_2 (DuPont Co.) as a white pigment; Dow Corning-11 (silicone glycol copolymer, Dow Corning Co.) as flowing agent; Byk-320 (Byk-Chemie GmbH Co.) as dispersant; Tinuvin-384 (benzotriazole derivative, Ciba-Geigy Co.) as UV absorber; Tinuvin-292 (HALS, Ciba-Geigy Co.) as UV stabilizer; and di-*n*-butyltindilaurate (DBTDL; reagent grade, Wako Pure Chemical Co.) as curing catalyst.

Synthesis of Pyrophosphoric Lactone Modified Polyesters

The synthetic conditions for TBOP [tetramethylene bis(orthophosphate)] intermediate are listed in Table I. The TBOP intermediate was prepared as follows: PYPA was poured in a 1L four-necked flask, and the temperature was raised from 36°C to 60°C for 90 min while slowly dropping BD into the flask, then it was held at 65°Cfor 2 h to complete the reaction. The product was purified by treatment with an excess of ethylether seven times to remove

TableI Synthetic Conditions for Lactone Modified Polyester, TBOP, and Pyrophosphoric Lactone Modified Polyesters

Products		TAP-2	TBOP	TAPT -10C	TAPT -20C	TAPT -30C
Reactants	TMP[a] g(mol)	142.1 (1.05)	-	143.4	144.7	146.0
	AA[b] g(mol)	157.3 (1.08)	-	127.2	97.2	67.1
	PCP[c] g(mol)	139.4 (0.26)	-	140.7	142.0	143.2
	TBOP[d] g	-	-	28.1	56.2	84.3
	PYPA[e] g(mol)	-	500.0 (2.81)	-	-	-
	BD[f] g(mol)	-	126.4 (1.40)	-	-	-
Toluene g		16	-	16	16	16
Reactions	Temp. (°C)	140~200	65	140~210	140~200	140~200
	Time (h)	13	2	8	7	7
Dehydration (mL)		38.0	Phosphoric Acid 273.5	39.0	39.8	40.4
Yield (%)		86	53	88	90	90

[a]TMP; Trimethylolpropane, [b]AA; Adipic acid, [c]PCP; Polycaprolactone 0201,
[d]TBOP; Tetramethylene bis(orthophosphate), [e]PYPA; Pyrophosphoric acid,
[f]BD; 1,4-Butanediol.

phosphoric acid, followed by vacuum drying at 5 mmHg and 40°C to obtain the dark-brown transparent viscous TBOP intermediate.

The conditions for synthesis of pyrophosphoric lactone modified polyester containing 10% PYPA are in Table I(TAPT-10C). The reactor was purged with N_2 gas at a flow rate of 30 ml/min at 70°C and the stirring rate was fixed at 250 rpm. The esterification reaction was carried out at a heating rate of 10°C/h. Dehydration started at 150°C and continued to 190°C. The reaction temperature was held constant at 190°C for 5 h and at 200°C for 6 h. The end point of reaction was determined by the amount of dehydration and the acid value. Unreacted reactants were removed with excess distilled water and methanol; leaving the brown transparent viscous TAPT-10C (pyrophosphoric lactone modified polyester containing 10% PYPA). The conditions for TAPT-20C and TAPT-30C, containing 20% and 30% PYPA respectively, are listed in Table I (TAPT-20C and TAPT-30C). The phosphorus content in TBOP intermediate was measured by the phosphomolybdate method(6) using UV spectroscopy.

Preparation of the Flame-Retardant Coatings

A modified polyester resin solution was prepared by diluting 100 g of TAPT (synthesized above) with 25.4 g of ethyl cellosolve, 25.4g n-butyl acetate, and 25.4g ethyl acetate respectively then blending with 90 g of TiO_2, 3.0 g of Byk-320, 1.0 g of Dow Corning-11, 1.0 g of Tinuvin-384, 0.5 g of Tinuvin-292, and 0.5 g of DBTDL. The isocyanate curing solution was prepared by mixing 136.1 g of Desmodur Z-4470, 15 g of ethyl cellosolve, and 15 g of xylene. Two-component polyurethane flame-retardant coatings was prepared by blending 272.7 g of the prepared modified polyester resin solution and 166.1 g of the isocyanate curing solution. The two-component polyurethane flame-retardant coating prepared from TAPT-10C/Z-4470, TAPT-20C/Z-4470, and TAPT-30C/Z-4470 were designated as TAPPU-10C, TAPPU-20C, and TAPPU-30C, respectively. The product of a blank experiment with TAP-2/Z-4470 was designated as TAPU-2. Three substrate materials, carbon steel, tin, and glass sheet were used to evaluate the physical properties of the coatings. Film thickness was controlled to be 0.076 mm with Doctor film applicator.

Characterization of Coatings

Infrared spectra were obtained using a Bio-Rad FTS-40 infrared spectrometer and ^{1}H-NMR spectra were obtained on a Varian Unity-300 system. Molecular weight was measured by GPC (R-410, Waters Co.) with 4 columns of Shodex KF-802, KF-803, KF-804, and KF-805 in series. Thermogravimetric analysis was carried out on a TGA-50H (Shimadzu, Japan).

To examine the flame retardancy of the flame-retardant coatings, three flame retardancy tests were performed. The 45° Meckel burner method(7) was used according to JIS Z-2150. Flame retardancy test conditions for various textiles were as follows: Textiles used were refined acrylic textile [100%, Ne 2/36, Hanil Co., Korea], nylon taffeta [70D/24F, Dongyang Nylon Co., Korea], and polyester taffeta [75D/24F, Samyoung Textile Co., Korea]; wet pick-up(8) condition was 80% acrylic textile, 60% nylon taffeta, and 60% polyester taffeta; and curing took place at 100°C for 5 min. The limiting oxygen index test (LOI method)(9) was performed using a flammability tester [ON-1, Suga Co., Japan], and the Cone calorimeter(10) was supplied by Fire Testing Technology Co., UK.

RESULTS AND DISCUSSION

Identification of the Synthesized Lactone Modified Polyester

In this study, the lactone modified polyester containing hydroxyl group was synthesized by the methods of Shoemaker(11), using PCP as diol, TMP as triol, and AA as di-basic acid. We could verify the existence of the ester group by C=O stretching vibration, C-O- stretching vibration, and hydroxyl group by OH stretching vibration using FTIR spectra and ^{1}H-NMR spectra. The polydispersity index is 4.08, and this is larger than general polydispersity index values. This indicates that side reactions occur, because the polyol monomer TMP has three functional groups.

Identification of Synthesized Pyrophosphoric Lactone Modified Polyesters

The reaction intermediate TBOP was synthesized prior to the synthesis of the lactone modified polyesters. Scheme 1 shows the synthesis of TBOP(12,13). The content of phosphorus in TBOP was 23.9%. The reaction scheme for the formation of TAPT is shown in Scheme 2. The presence of phosphorus is shown by the P-O-C stretching vibration at 1020 cm^{-1} and P=O stretching vibration at 1240 cm^{-1}. Table II shows the GPC measurement results of TAPT-10C and TAPT-30C; the molecular weights increase with the content of PYPA.

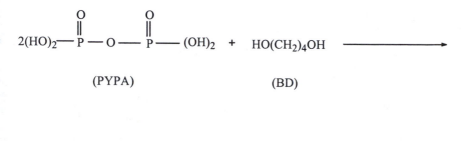

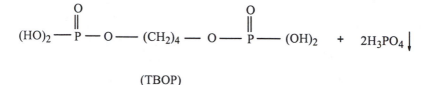

Scheme 1. Synthesis of TBOP.

Table II. Molecular Weight of TAP-2 and TAPT Resins[a]

Resin Code	M_n	M_w	M_z	M_w/M_n
TAP-2	2600	10600	37500	4.08
TAPT-10C	2900	14000	49300	4.83
TAPT-30C	3300	26100	149100	7.91

[a]Measured from GPC run on μ-Stragel columns with THF as eluent.

Thermal Stability

It is known that the flame-retardant action of phosphorus is caused not only by the formation of a char but also by the expansion of the thick porous char layers produced(15). Because of this action of phosphorus, it is possible to achieve high flame retardancy for organic compounds containing oxygen, and moreover, it is possible to take advantage of afterglow prevention by phosphorus.

116

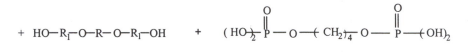

(TMP)   (AA)

+ HO—R₁—O—R—O—R₁—OH + (see structure)

(PCP)   (TBOP)

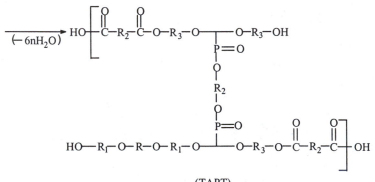

(TAPT)

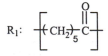

R_1: $-(CH_2)_5-C(=O)-$

R_2: $-(CH_2)_4-$

R_3: $-CH_2-C(CH_3)(CH_2)(CH_2OH)-CH_2-$

Scheme 2. Synthesis of TAPT.

Figure 1 shows the results of thermal analysis of TAPTs containing phosphorus. TAPT-10C and TAPT-30C showed 36.90 and 79.24% weight losses, respectively, at 400°C, and residues were 10 and 19% at 600°C. TAPT-10C decomposed suddenly at 400°C and TAPT-30C showed large weight loss at 330°C. It was found that the decomposition temperature was decreased with an increase of phosphorus content. With the increase of phosphorus content, the decomposition temperature was lowered and the amount of final char increased rapidly.

The introduction of phosphorus moiety into polyurethane leads to a decrease in the onset temperature of degradation and an acceleration of char formation at higher temperatures(16). Lawton and Setzer(17) discovered that compounds containing halogen or phosphorus, added before melt spinning PET textile, revealed a flame-retardant property. Koch et al.(18) also reported that the thermal decomposition of PET textile containing phosphorus took place at lower temperature.

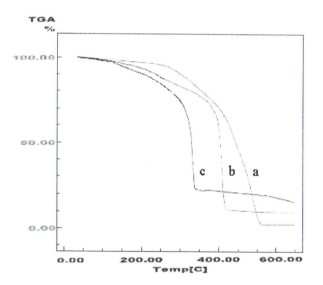

*Figure 1 TGA thermogram of (a) TAP-2, (b) TAPT-10C,
and (C) TAPT-30C.*

Comparison of Flame Retardancy of Coatings

The flame-retardant coatings do not burn vigorously and stop burning spontaneously upon removal of flame. In a flame-retardant coating, halogen acts as a flame-retarding component through the prevention of the flaming combustion of decomposition gases where the non-toxic phosphorus acts as a flame-retardant component through the prevention of the non-flaming combustion of decomposition residues(19). During the burning, phosphorus compound is converted to phosphoric acid, meta-phosphoric acid, and poly-meta-phosphoric acid; these acids, produced on the film of the coating, block the heat and oxygen flow, and thereby phosphorus acts as a flame-retardant component(20). Phosphorus compounds show outstanding flame-retarding action through two different mechanisms(21).

Table III. Flame Retardancy of Fabrics Treated with Flame Retardant Coatings

	Exp. No	Concent-ration (wt%)	45°Meckel burner method		
			Char length (cm)	Afterflaming[a] (sec)	Afterglow[b] (sec)
Acrylic Fabrics	Untreated	-	BEL[c]	-	-
	TAPPU	10	18.5	23.0	1.0
		20	4.5	0	0
		30	3.9	0	0
Nylon Taffeta	-	-	8.9	0	0
	B-2[b]	-	BEL	-	-
	TAPPU	10	3.9	0	0
		20	3.3	0	0
		30	3.2	0	0
Polyester Taffeta	B-1[a]	-	7.9	2	0
	B-2[b]	-	BEL	-	-
	TAPPU	10	3.6	0	0
		20	3.4	0	0
		30	3.1	0	0

[a]Original fiber not treated with flame retardant coating and resin.
[b]Fiber treated with resin only.
[c]Burned entire length.

Flame retardancy results with the 45° Meckel burner method are listed in Table III. The char lengths of acrylic fabrics treated with 20wt% and 30wt% TAPPU were both below 4.5 cm; afterflaming and afterglow were both below 1 sec. These results show that the flame-retardant coatings on acrylic fabrics are excellent and are good flame retardant coatings for acrylic fabrics. The results for nylon taffeta and polyester taffeta are similar to the results for acrylic fabrics.

Figure 2 shows LOI values as a function of PYPA content. The flame retardant effect increases with PYPA content; the LOI value was 30% when PYPA content was 30wt%. Since the LOI values of typical non-flame-retardant coatings are around 17%, the flame retardancy was much improved.

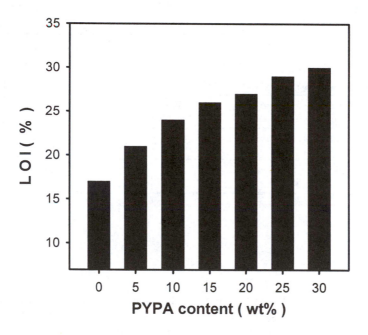

Figure 2 Relationship between LOI and pyrophosphoric acid content of lactone modified polyester in the two-component polyurethane flame retardant coatings.

Casu et al.(22) used heat release rate (HRR) from the Cone calorimeter to compare the flame retardancy of nylon 66 and nylon 6 treated with melamine cyanurate. Table IV lists testing conditions and the peak-HRR, average-HRR, and average-EHC(average effective heat of combustion). The peak-HRR values, maximum heat produced by burning, decrease with on increase of phosphorus content.

A graphical representation of the peak-HRR is shown in Figure 3. The values decrease with as phosphorus increases, and the time of peak-RHR is also shifted to the longer times.

Table IV. Fire Test Results Obtained from Cone Calorimeter

Item	Sample	TAPU-2	TAPPU-10C	TAPPU-30C
Specimen thickness (mm)		2.0	2.0	2.0
Heat flex (kW/m^2)		50	50	50
Exhaust duct flow rate (m^3/sec)		0.024	0.024	0.024
Peak-heat release rate (kW/m^2)		1090	854	711
Average-heat release rate (kW/m^2)		420	361	318
Average-heat combustion (MJ/m^2)		26.2	26.0	22.4

Conclusions

The use of phosphorus-containing polyurethane flame retardant coatings enhances the thermal stability, relative to the non-phosphorus containing material. Two mechanisms, a decrease in the onset temperature of degradation and an acceleration of char formation at higher temperatures, are suggested to explain this reason.

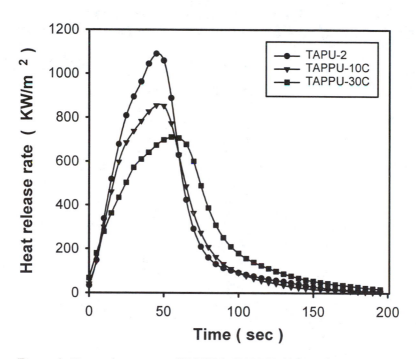

Figure 3 Heat release rate of TAPU-2, TAPPU-10C, and TAPPU-30C.

Acknowledgements

This work was financially supported by the Korean Ministry of Commerce, Industries and Energy through the grant for Clean Production Technology (99-1-M-11-4) and also supported by the Korean Ministry of Education through BK21 Program.

References

1. Im, W. B.; Park, H. S. *J. Kor. Oil Chem. Soc.*, **1998**, *15*, 77-84.
2. Ma, Z. L.; Zhao, W. G. ; Liu, Y. F.; Shi, J. R. *J. Appl. Polym. Sci.*, **1997**, *66*, 471-475.
3. Liu, Y. L.; Hsiue, G. H.; Lan, C. W.; Chiu, Y. S. *J. Polym. Sci.(Part A)*, **1997**, *35*, 1769-1780.
4. Weil, E.; McSwigan, B. *J. Coat. Technol.*, **1994**, *66(839)*, 75-82.
5. Kordomenos, P. I. ; Frisch, K. C.; Xiao, H. X.; Sabbah, N. *J. Coat. Technol.*, **1985**, *57(723)*, 23-28.
6. Yoo, H. J.; Lee, H. J. *J. Kor. Fiber Soc.*, **1997**, *34*, 451-458.
7. Park, H. S.; Kim, Y. G.; Bea, J. S. *J. Kor. Oil Chem. Soc.*, **1990**, *7*, 31-36.
8. Park, H. S.; Kim, Y. G.; Kim, J. T.; Bea, J. S.; Choi, H. C. *Polymer(Korea)*, **1990**, *14*, 211-219.
9. Hindersinn, R. R.; Witschard, G. *Flame Retardancy of Polymeric Materials;* Marcel Dekker, Inc.: New York, 1982; Vol. 4, pp 6-8.
10. Im, W. B. Ph. D. thesis, Myongji Univ., Yongin, Korea, 1998.
11. Shoemaker, S. H. *J. Coat. Technol.*, **1990**, *62(787)* , 49-55.
12. Zech, J. D.; Ford, E. C. *U. S. Patent* 3,309,427, **1967**.
13. Wismer, M.; Poerge, H. P.; Mosso, P. R.; Foote, J. F. *U. S. Patent* 3,407,150, **1968**.
14. Li, T.; Graham, J. C. *J. Coat. Technol.*, **1993**, *65(821)*, 63-69.
15. Roth, S. H.; Green, J. *J. Paint Technol.*, **1974**, *46*, 58-62.
16. Park, H. S. *J. Kor. Fiber Soc.*, **1997**, *34*, 386-392.
17. Lawton, E. L.; Setzer, C. J. *Flame Retard. Polym. Mater.*, **1975**, *1*, 193-199.
18. Koch, P. J.; Pearce, E. M. J.; Lampham, A.; Shalaby, S. W. *J. Appl. Polym. Sci.*, **1975**, *19*, 227-233.
19. Papa, A. J. *Flame Retardancy of Polymeric Materials*; Marcel Dekker, Inc.: Papa, A. J. Papa, A. J. New York, 1980; Vol. 3, pp 1-61.
20. Park, H. S.; Kwon, S. Y.; Seo, K. J.; Im, W. B.; Wu, J. P.; Kim, S. K. *J. Coat. Technol.*, **1999**, *71(899)*, 59-65.
21. Keun, J. H.; Park, H. S. *J. Kor. Oil Chem. Soc.*, **1994**, *11*, 45-52.
22. Casu, A.; Camino, G.; De Giorgi, M.; Flath, D.; Morone, V.; Zenoni, R. *Polym. Degrad. Stab.,* **1997**, *58*, 297-302.

Non-Halogen Fire Retardants

The most traditional area of fire retardancy studies involves additives to enhance the thermal stability of a polymer and render it fire retardant. It is thus not surprising that the largest number of papers in this book deal with such additives. There has been considerable concern in the environmental community about the use of halogen additives. That concern is based upon the perceived potential for halogenated dioxin and furan formation during a fire and the perceived lack of recyclability of resins containing halogen flame retardants. Whether or not those concerns are scientifically valid, there has been a vigorous effort in the area of non-halogen approaches to flame retardancy. There are many fruitful approaches. Thus the topics that are covered here, representative of the state of the art, are quite varied.

The initial chapter provides data on a novel cross-linking system in which Friedel-Crafts chemistry is used to enhance the thermal stability of polystyrene. The heat release arte is significantly reduced and the polymer, when combined with the Friedel-Crafts catalyst and cross-linking agent, shows and onset temperature of degradation which is at least 50°C higher than that of the pure polymer.

One of the most promising additive approaches is intumescence and this is represented in two papers, describing the additives that are used and the characterization of an intumescent system. Some of the traditional additives which are covered include phosphorus compounds, metal hydroxides and zinc borates.

Another important area is the processing of the polymer and additives. This is a new area for this series and two papers describe processing and its effect on the efficacy of fire retardant systems.

The other two papers describe quite novel systems. One of these uses a cold remote plasma process to deposit organosilicon thin films on a polymer and this shows improvement in oxygen index and in heat release rate. The last paper in this section presents a novel system for polyester fabric.

Chapter 10

Cross-Linking of Polystyrene by Friedel–Crafts Chemistry: A Review

Hongyang Yao and Charles A. Wilkie

Department of Chemistry, Marquette University, P.O. Box 1881, Milwaukee, WI 53201-1881

The cross-linking of polystyrene by Friedel-Crafts chemistry is briefly reviewed. The focus has been on producing a cross-linked structure which will char and provide thermal protection to the underlying polymer. The mechanism and kinetics of cross-linking along with the identity and activity of the functional groups have been of interest. The key parameter is the cross-linking temperature, control of which could be achieved through the choice of functional groups, catalysts, and inhibitors. The utilization of Friedel-Crafts chemistry to promote thermal stabilization and flame retardancy for styrenic polymers has been the intent of this study.

The annual production of polystyrene and styrenic copolymers (ABS, SBS, etc.) in North America is about 9.7billion lb in 1999 and the major applications are to make polystyrene, HIPS, ABS, SBS (1). These polymers are widely used in toys, housewares, home appliances, furniture, and automotive parts. Like most commodity polymeric materials, styrenics are intrinsically flammable, *i.e.*, vulnerable to rapid oxidative thermal degradation. The burning involves the degradation of bulk polymers to volatile products, whose combustion provides heat which is fed back to the surface of the polymer to continue the degradation and generate more energy to break additional bonds in the polymer chains.

In order to achieve flame retardancy, one can utilize one or more of the following strategies: 1. Inhibition of the gas phase reactions by free radical traps; 2. Modification of the pathway of pyrolytic decomposition of the bulk polymer to remove or reduce the fuel source; and 3. formation of a barrier which prevents the transfer of the energy of combustion to the polymer and thus prevents further degradation. This last pathway may involve either the premixing of an inorganic filler or the *in situ* formation of a char layer. Inorganic-polymer blends have been widely used, and the recent advance of clay-polymer nanocomposites provides a new dimension to this method (2). The char generation *in situ* is through reactions of polymer with itself or with appropriate additives, which can be promoted by graft copolymerization, Lewis acid catalysis, Friedel-Crafts chemistry, redox reactions, etc (3).

Most commercial flame retardants for styrenics contain bromine/chlorine compounds along with metal oxides synergists. In pursuit of alternatives to these systems, cross-linking reactions of polymers have been explored as a strategy to promote the formation of char and thus enhance the thermal stability and lower the flammability (4,5,6). The phenyl rings in polystyrene are in pendant positions, and the major reaction pathway is random scission, which leads to the formation of monomer and oligomers. In order to form char, the phenyl rings must be connected to form graphitic-like structures before the chain scission occurs. In any study of flame retardancy, one must be concerned with a variety of parameters, such as ease of ignition, flame spread, heat release rate, ease of extinction, fire endurance, smoke and toxicity (7). Recent advances on the cross-linking of polystyrene by Friedel-Crafts reactions are reviewed in this paper.

Cross-linking of Polystyrene by Friedel-Crafts Reaction

The Friedel-Crafts reaction of interest for this study is the same reaction used in organic chemistry in which an alkyl halide, or other alkylating agent, is added to an aromatic ring in the presence of a suitable catalyst, typically aluminum trichloride for the conventional organic system (8). Since this reaction normally proceeds quite easily at room temperature for simple organic molecules and since $AlCl_3$ is not a desirable material to put into a polymer, a few variations are necessary. The catalyst and/or the alkylating agent must be deactivated so that reaction does not occur below the processing temperature. The desired reaction should utilize a difunctional alkylating reagent, so that cross-linking rather than simple alkylation can occur, in the presence of a suitable catalyst at a temperature somewhat above 200°C. The desired structures to be formed should be more stable than the main chain sp^3-sp^3 C-C bonds, so that the polymer fragments will be retained in the condensed phase and form a graphitic-like network.

The Friedel-Crafts reaction has been successfully utilized to cross-link polystyrene with metal chlorides as catalyst several decades ago; the cross-linking temperature is too low and the resulting polymer has a lower thermal stability than the virgin polymer (9,10,11). In the initial work from this laboratory, a difunctional alkylating agent, 1,4-benzenedimethanol, was found to cross-link polystyrene at 300°C, but not at 200°C, in a sealed tube, catalyzed by various zeolites (12). The resulting cross-linked polymers were significantly more thermally stable than the virgin polystyrene. Unfortunately the diol volatilizes before it can react when the combination of polystyrene, diol, and zeolite are reacted in a flowing system. In order to make this chemistry useful, it is necessary for both the alkylating agent and catalyst to be available for the reaction.

Incorporation of Functional Groups by Copolymerization

Structural modification of an inherently flammable polymer is one way to make it less flammable or respond better to flame retardants. Pearce showed that the introduction of chloromethyl groups on the aromatic rings by copolymerization with vinylbenzyl chloride, along with the addition of antimony oxide or zinc oxide to provide a latent Friedel-Crafts catalyst, could increase the char formation of polystyrene (13,14,15).

Functional groups can be incorporated into polystyrene by copolymerization of styrene with vinylbenzyl chloride or vinylbenzyl alcohol (16,17,18). As expected, the chloride-containing copolymer cross-links more easily than alcohol-containing material when metal halides are used as the catalyst. These metal halides are very effective catalysts for Friedel-Crafts chemistry and cause cross-linking to occur at temperatures below the processing temperature of polystyrene and hence are unsuitable for our purpose.

An effective catalyst is 2-ethylhexyldiphenylphosphate, DPP; this material decomposes at about 220°C by the loss of 2-ethylhexene to give the corresponding phosphoric acid which is the effective catalyst (16). Thus cross-linking is controlled since the catalyst is not present until the ester undergoes degradation to form the acid, as shown in Figure 1.

The combination of the alcohol-containing copolymer with DPP has been studied by TGA/FTIR, Cone calorimetry and radiative gasification. A schematic diagram of the chemical reaction which occurs to give rise to cross-linking is shown in Figure 2. Figure 3 shows the TGA/FTIR study of the copolymer with DPP. The degradation of DPP causes the evolution of aromatics at about 250°C, and thus the earlier onset of the degradation from the mixture is due to the degradation of the DPP, but the principal evolution of aromatics, due to the degradation of the polymer, commences at a higher temperature than for the virgin polymer.

128

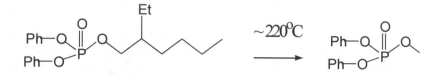

Figure 1. Degradation of 2-ethylhexyldiphenylphosphate, DPP, to give a phosphoric acid.

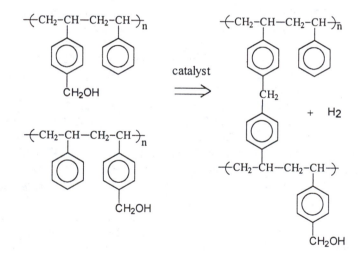

Figure 2. Schematic of the cross-linking reaction of functionalized polystyrene.

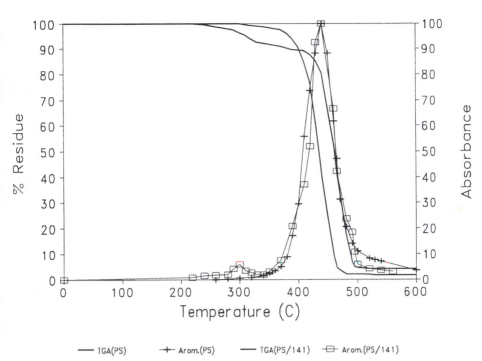

*Figure 3. Thermogravimetric analysis and absorbance of evolved aromatics
from polystyrene and a blend of polystyrene with 5% DPP(i.e., 141)*

The results of Cone calorimetry show that the peak rate of heat release is significantly reduced for the efficacious combination relative to the various controls; this is shown graphically in Figure 4. Radiative gasification experiments show that a char is formed on the surface of the polymer and it is inferred that this does insulate the underlying polymer from the heat source. Figure 5 shows the mass loss rate as well as the temperature of the thermocouple beneath the sample. The observation of a lower temperature beneath the sample for the efficacious combination shows the insulating ability of the char which is formed. The reduction in mass loss rate also shows the effect of this char(17).

Various concentration of alcohol group have been studied and different number (n = 1, 2, 3) of methylene groups between the aromatic ring and the alcohol functionality have been synthesized, and their effects on the cross-linking and thermal stability were studied by gel-content, swelling ratio and TGA (18); the structure of the copolymers is shown below as Figure 6. There is little difference between the three copolymers in terms of thermal stability. An amount as small as 3% of the alcohol containing fraction produces a system with a large gel content and a high cross-link density.

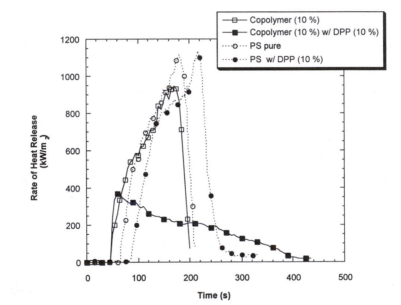

Figure 4. Rate of heat release for pure PS, PS with DPP, poly(styrene-co-4-vinylbenzyl alcohol) and poly(styrene-co-4-vinylbenzyl alcohol) with DPP at a heat flux of 35 kW/m² (reprinted from Polymer Degradation & Stability, volume 99, Authors: Z. Wang, D. D. Jiang, C. A. Wilkie, and J. W. Gilman, pp. 373-378, Copyright (1999), with permission from Elsevier Science).

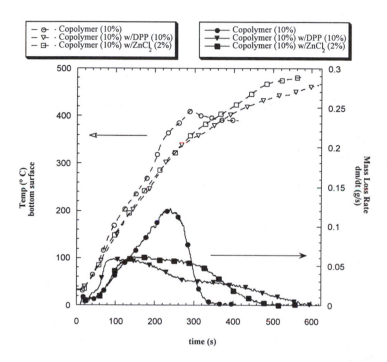

Figure 5. Mass loss rate and bottom surface thermocouple data from gasification experiments performed at a heat flux of 40 kW/m², in nitrogen for poly(styrene-co-4-vinylbenzyl alcohol) copolymer with 2-ethylhexyldiphenylphosphate, DPP (DPP mass fraction 10%; 4-vinylbenzyl alcohol mass fraction 10%) and with zinc chloride (ZnCl₂ mass fraction 2%) (reprinted from Polymer Degradation & Stability, volume 99, Authors: Z. Wang, D. D. Jiang, C. A. Wilkie, and J. W. Gilman, pp. 373-378, Copyright (1999), with permission from Elsevier Science).

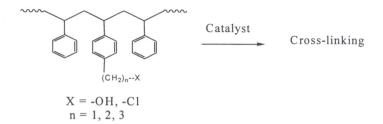

X = -OH, -Cl
n = 1, 2, 3

Figure 6. Cross-linking of styrene copolymers with a variation of the copolymer composition.

Multi-functional Additives

Based upon these observations, one may conclude that this is a very effective system to lower the heat release rate of polystyrene. One significant liability is the need to use a copolymer, which is not commercially available, rather than the commercial, and widely used, polystyrene. In an effort to increase the utility of this system, we have returned to the original idea of a difunctional additive to be reacted with polystyrene. However, the intent is now to have two different functionalities which will react at different temperatures. In the first step, a functionalized polymer will be produced and this reaction should occur under processing conditions. In the second step, cross-linking of the functionalized polystyrene occurs, leading to thermal stabilization of the polymer.

The critical parameter is the temperature of the cross-linking (T_c). It must be higher than the processing temperature ($T_p \sim 200°C$) to be practical and lower than the degradation temperature ($T_d \sim 350°C$) to be useful. But in all previous attempts to utilize Friedel-Crafts chemistry, T_c is lower than T_p, so our efforts were devoted to raising T_c. The desired process, both in terms of the difunctional additive and the temperature requirements, is shown in Figure 5.

Various multi-functional compounds have been tested and some functional groups undergo reaction too easily while others do not react under the required conditions. In the category of those which cannot effect cross-linking are placed diols, diethers, dialdehydes and combinations of these functionalities along with a vinyl group and one of these functionalities. Species which will cause cross-linking under any temperature conditions include vinyl-phosphate , vinyl-tosylate, ditosylates, dimesylates, and diphosphates. The preferred category, those which will cause cross-linking only under the preferred temperature conditions include vinylbenzyl alcohol, vinylbenzylnitrate, dichloroxylene, and combinations of benzyl alcohol and benzyl chlorides. A particularly effective agent is *ar*-hydroxymethylbenzyl chloride (HMBC) (19). The *para* isomer, *p*-HMBC, can cross-link polystyrene at a temperature as low as 150°C with a concentration as low as 0.1 wt % and without a catalyst. The cross-linking

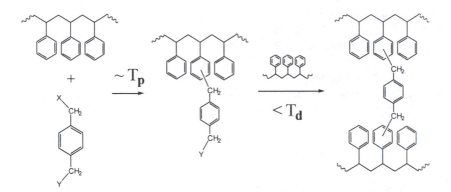

Figure 7. Cross-linking of polystyrene using a difunctional additive and its temperature requirements.

temperature can be controlled by the addition of catalysts and inhibitors, to be in between 200 and 250°C.

Catalysts and Inhibitors

The Friedel-Crafts reaction can be catalyzed by almost any acid, such as metal chlorides, yet these are not appropriate in this instance, due to their high efficiency at even very low temperature. The preferred catalyst is DPP, which has been described above. If the difunctional additive contains a halide, then hydrogen halide will be evolved and it would be advantageous to remove this so as to prevent damage to the processing equipment, and this implies the presence of some base. The choice of the base is important, since bases can inhibit the Friedel-Crafts reaction, unless the base can function only as an acid acceptor and not as a nucleophile. Most amines, such as 1,8-bis(dimethylamino)-naphthalene, 1,8-diaminooctane, are so active that they completely inhibit the cross-linking reaction until a temperature higher than 300°C. The hindered amine (HALS), 2,2,6,6-tetramethyl-4-piperidinol (TMP), was found to be less active. The amount has been optimized to about 0.1 wt% and effective up to about 220°C, until it was consumed by the resultant phosphoric acid from the phosphate degradation.

When the combination of polystyrene, DPP as catalyst and p(HMBC) as the difunctional additive was studied using TGA/FTIR, it was found that both hydrogen chloride and water were evolved at the same time. This surprising result means that both the alcohol and the chloride functionality are equally reactive under processing conditions but, once one end has reacted, the other will not react until significantly higher temperatures. Fire performance was evaluated using Cone calorimetry. The most efficacious system used 10% of the difunctional additive, 5% DPP and 0.1% of a hindered amine, 2,2,6,6,-

tetramethyl-4-piperidinol. This system reduced the peak heat release rate to 344 kW/m^2, a reduction of 68% (20).

While the reduction in the rate of heat release is substantial, the utilization of Friedel-Crafts chemistry alone can not give a useful result in the Underwriter's Laboratory, UL-94, protocol. It is worthwhile to note that a synergistic combination between conventional vapor phase flame retardants, such as halogens or phosphorus compounds, will give a V-0 in this protocol(20).

Conclusions

It is known that the cross-linking of polymers is a strategy to promote the formation of char, and thus results in enhanced thermal stability and flame retardancy. The efforts to use of Friedel-Crafts chemistry to cross-link polystyrene has been reviewed as one promising way to achieve flame retardancy. The mechanism and kinetic aspects were also discussed in order to understand the role of each component and provide foundation for the formulation of flame retardant.

Acknowledgment

This work was partially supported by the US Department of Commerce, National Institute of Standards and Technology, grant number 60NANB6D0119, and the Albemarle Corporation.

References

1. *C&EN*, June 26, 2000.
2. Gilman, J. W.; Kashiwagi, T.; Giannelis, E. P.; Manias, E.; Lomakin, S.; Lichtenhan, J. D.; Jones, P. in *Fire Retardancy of Polymers The use of intumescence*, Royal Soc special Publ 224, Le Bras, M.; Camino, G.; Bourbigot, S.; Delobel, R. Eds., Royal Society, Cambridge, 1998, pp 203-221, and references cited therein.
3. Levchik, S.; Wilkie, C.A. in *Fire Retardancy of Polymeric Materials,* Grand, A.F.; Wilkie,C.A. Eds., Marcel Dekker, NY, 2000, pp 171-215.
4. Van Krevelen, D. W. *Polymer,* **1975,** 16, 615-620.
5. Lyon, M. *ANTEC'96 Processing*, paper 538, Indianapolis, IN, May 1996, p.3050.
6. Nyden, M.; Coley, T. R.; Mumby, S. *ANTEC'96 Processing,* paper 540, Indianapolis, IN, May 1996, p.3058.

7. Nelson, G. L. in *Fire and Polymers II,* Nelson, G. L., Ed., ACS Symposium Series 599, American Chemical Society, Washington, DC, 1989; p 1-26.
8. *Friedel-Crafts and Related Reactions,* G. A. Olah, Ed.; Wiley-Interscience, New York, 1964.
9. Grassie, N.; Gilks, J. *J. Polym. Sci.: Chem. Ed.,* **1973,** 11, 1985-1994.
10. Brauman, S. K. *J. Polym. Sci.:Polym. Chem. Ed.,* **1979,** 17, 1129-1144.
11. Rabek, J. F.; Lucki, J. *J. Polym. Sci.: Part A: Polym. Chem.,* **1988,** 26, 2537-2551.
12. Li, J., Wilkie, C. A., *Polym. Degrad. Stab.,* **1997,** 57, 293-299.
13. Khanna, Y. P.; Pearce, E. M. in *Flame Retardant Polymeric Materials,* Lewin, M.; Atlas, S.; Pearce, E. M.; Eds.; Vol. 2, Plenum Press, NY, 1975; pp. 43-61.
14. Pearce, E. M. *Contemp. Topics in Polym. Sci.,* Vol. 5; Vandenberg, E. J., Ed.; Plenum Press, NY, **1984;** pp. 401-413.
15. Pearce, E. M. in *Proceedings of the Meeting on Recent Advances in Flame Retardancy of Polymeric Materials,* Lewin, M., Ed., CT, 1991, pp. 116-120.
16. Wang, Z.; Jiang, D. D.; McKinney, M. A.; and Wilkie, C. A. *Polym. Degrad. Stab.,* **1999,** 64, 387-395.
17. Wang, Z.; Jiang, D. D.; Wilkie, C. A.; Gilman, J. W. *Polym. Degrad. Stab.,* **1999,** 66, 373-378.
18. Zhu, J.; McKinney, M. A.; Wilkie, C. A. *Polym. Degrad. Stab.,* **1999,** 66, 213-220.
19. Yao, H.; Zhu, J.; McKinney, M. A.; Wilkie, C. A. *J. Vinyl. Add. Tech.,* in press
20. Yao, H; McKinney, M. A;, Dick, C.; Liggat, J.J.; Snape, C.E.; Wilkie, C. A. manuscript in preparation.

Chapter 11

Use of Carbonizing Polymers as Additives in Intumescent Polymer Blends

M. Le Bras and S. Bourbigot

Laboratoire de Génie des Procédés d'Interactions Fluides Réactifs–Matériaux, E.N.S.C.L., BP 108, F-59652 Villeneuve d'Ascq Cédex, France

Intumescence is one way to obtain effective halogen-free fire retarded polymeric systems. The first generation of intumescent systems consisted of mixtures of polyols as carbon sources together with sources of acid, typically ammonium polyphosphate, APP. The development of the corresponding materials is limited by low compatibility between the host matrix and additives, eventual reaction between the additives during processing or polyols hydrolysis and subsequent exudation. A solution consists in the substitution of the polyols by polymers which undergo a natural carbonization upon heating: thermoplastic polyurethanes, polyamide-6 or clay-polyamide-6 nanocomposites. This study shows that mixtures of these polymers with ammonium polyphosphate are effective fire retardant additives in EPR, PP, PS and EVA. Finally, it is shown that FR performances of polyamide-6 based material are preserved after an artificial aging.

Fire protection of flammable materials via an intumescence process has been known for several years: intumescent materials on heating form foamed cellular charred layers on the surface. These layers act as a physical barrier which slows down heat and mass transfer between the gas and condensed phase.

Generally, intumescent formulations contain three active ingredients: an acid source (such as ammonium polyphosphate (APP) used in this work), a carbon source and a blowing agent. The first generation of carbon sources used in intumescent formulations for thermoplastics consisted of polyols such as pentaerythritol, mannitol or sorbitol (1, 2). Problems with this kind of additives consist in migration/blooming of the additives, water solubility of the additives and reaction with the acid source during processing of the formulations (2-4). Moreover, there is not a good compatibility between the additives and the polymeric matrix and the mechanical properties of the polymer are comparatively poor.

Recently, we have developed fire retardant (FR) intumescent formulations using charring thermoplastic polymers (thermoplastic polyurethanes (TPU) (5), polyamide-6 (PA-6) (6) and hybrid clay-PA-6 nanocomposites (PA6nano) (7, 8)) as carbon sources. The advantage of the concept is to obtain FR polymers blends with improved mechanical properties and to avoid the problems of migration and solubility of the additives (9).

The purpose of this paper is to review the FR and mechanical performances of these formulations. Most particularly, the influence of the chemical structure of the TPU has been investigated by using two different TPU series, which differ in the nature of the polyol used for their syntheses: polyols in the polyaddition process are either polyether or polyester. Generally, a TPU synthesized from polyether shows hard segment domains (diol + diisocyanate) larger than those found in a polyester-based TPU. So, the influence of the hard segment content is studied in this work using different R ((diol + diisocyanate/ polyol) = R). Moreover, it is shown that substituting PA-6 by PA6nano leads to an increase of the FR and mechanical properties of an intumescent ethylene – vinyl acetate copolymer (EVA24 with 24 wt.% of vinyl acetate)-based material materials. Finally, this paper reports a recent study of properties of the EVA24/APP/PA-6 formulation after its artificial aging using Xenotest.

Experimental

Materials

Raw materials were isotactic polypropylene (PP, Eltex P HV219, as pellets supplied by Solvay), ethylene-propylene rubber (EPR, Hifax 70-36 as pellets supplied by Himont), several ethylene –vinyl acetate copolymers (EVAx with x = wt.% of vinyl acetate, Evatane/Lacqtène as pellets supplied by Elf Atochem), polystyrene (PS; crystal grade as pellets supplied by Elf Atochem), "high impact" PS (HIPS; as pellets supplied by Elf Atochem), PA-6 (as pellets supplied by Nyltech), polyamide-6 clay nanocomposite (PA-6-nano, as pellets supplied by UBE, clay mass fraction of 2 wt.-%), APP (AP 422: $(NH_4PO_3)n$, $n \approx$ 700, powder supplied by Clariant) and several TPU (as pellets supplied by Elastogran BASF). TPU 1185A10 (A) is a polyether-based TPU. In the polyester-based TPU series: B90A10, C88A10, S90A10 (B, C and S90) the chemical nature of polyol is changed. From S85A10 (S85), S90 to S74D (S74) the number of hard segments increases (R increases). Additives were always incorporated at 40 wt.-% in the polymer. The study was carried out on formulations processed using a counter-rotating twin screw extruder in the standard conditions: average residence time from 20 to 30 seconds, rotation of screw: 80π rad.mn^{-1}, melting, mixing zone and die temperature depending of the host matrix (5, 9).

Methods

Fire testing. Limiting Oxygen Index (LOI) was measured using a Stanton Redcroft instrument on specimens (100x10x3 mm^3) according to ASTM 2863 (10). UL-94 testing were carried out on 127x12.7x3 mm^3 specimenss according to the UL-94 testing (11). Samples (100x100x3 mm3) were exposed to a Stanton Redcroft Cone Calorimeter under an external heat flux of 50 kW/m^2 (this external flux corresponds to the heat evolved during a well developed fire (12)). Three tests have been performed on each material and the averages are used that limit the uncertainties on the measurements (relative variations of the data generally lower than 10 %). The Rate of Heat Release (RHR) represents the evolution of energy flow versus time for a given sample surface. The data were computed using the oxygen consumption principle using software developed in our laboratory.

Mechanical properties. The mechanical properties of materials are evaluated at 20°C by tensile testing on an INSTRON machine. Nominal stress-strain curves are recorded. The dimensions of the samples are: useful length and width (without head): 22 x 5 mm^2. Samples are prepared from compression molded plates with 3 mm thickness. The constant cross-head speed (50mm/min) corresponds to a strain rate of about 4.1×10^{-2} s^{-1}. Stress, elongation at break and Young modulus measured at a lower speed (5 mm/min) are recorded.

Aging. Simulated accelerated weathering test are carried out on sheet under a xenon lamp exposure and rain and temperature cycling (13). Duration of the test (700 h) gives results which, as an approximation, may be compared with a 1 year natural weathering in a temperate zone climate (14).

Results and Discussion

Thermoplastic polyurethanes-based additives

FR performance

The addition of TPU/APP in isotactic PP increases the LOI values of the formulations in each case. The evolution of the LOI values of PP-based formulations versus APP:TPU ratio shows a synergistic effect (figure 1), which depends directly on the nature of the polyol; the maximum of synergy is observed for a 1:1 for B, 2:1 for A and C and 3:1 ratio for S. In the case of the S series, a significant increase of LOI is observed when R ratio increases. It may be noticed that the polyester-based TPU (B, C, S series) provides the best FR performance in comparison with the polyether-based TPU (A).

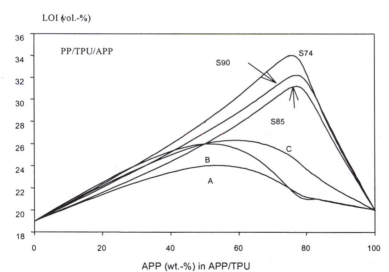

Figure 1: LOI values versus percentage of APP in PP/TPU/APP (Reproduced with the permission from reference 5. Copyright 1999 Technomic).

The formation of an expanded intumescent shield during UL 94 testing, and a time of combustion lower than 30 seconds are observed, but burning drops are always observed (V2 rating) with 40 wt.-% additive loading. V0 ratings are only obtained when the additive loading reaches 45 wt.-%.

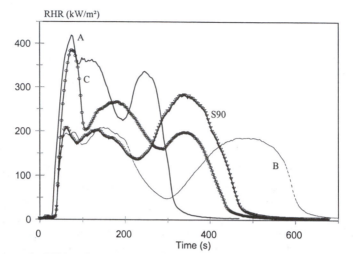

Figure 2: RHR values versus time for PP/APP/TPU (external heat flux: 50kW/m²); TPU A, B, C and S90 ((Reproduced with the permission from reference 5. Copyright 1999 Technomic).

The RHR curves (figures 2) of samples presenting the maximum synergistic effect according to LOI test show a significant decrease in RHR (RHR maximum values lower than 400 kW/m²) in comparison with the virgin polymer (RHR peak of virgin PP = 1600 kW/m²). The RHR evolution versus time including three steps is typical of an intumescent system (*15*). The best FR performances are obtained using polyester-based TPU and the nature of the polyol is important as illustrated by B which shows the lowest RHR peak (Figure 2) and low LOI (Figure1); RHR decreases when the number of hard segments increases (5).

FR properties of interest are also observed in EPR/TPU/APP formulations (*16*). As a typical example, 40 wt.-% loading of the S90/APP (4/1 wt./wt.) additive system in EPR gives LOI = 28 vol.-% and a maximum RHR about 400 kW/m².

In recent papers , the protective effect of intumescence is studied to determine the thermal stability of the FR PP/S90/APP and EPR/S90/APP blends under thermo-oxidative degradation conditions. Thermogravimetric analysis allows the computation of the degradation rate of a particular system (*16*, *17*). The study shows that a reaction between APP and TPU leads to the formation of an intumescent shield and gives a material with improved thermal stability; TPU acts as a classical intumescent carbonizing additive (18).

Mechanical performance

Evaluation of the mechanical performance of prototype materials is of great importance in order to determine their potential applications. We chose to characterize these materials mainly by their modulus, stress and elongation at break (Figure 3).

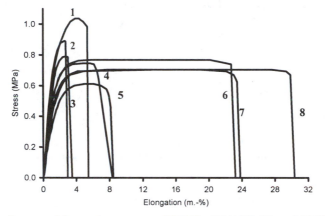

Figure 3: stress/elongation curves of PP(1), PP/APP (2) and TPU-based materials (S74 (3), S90 (4), B (5), S85(6), A (7) and C (8)) ((Reproduced with the permission from reference 5. Copyright 1999 Technomic).

The Young's modulus is generally lowered compared with extruded PP and they have the typical behavior of ductile materials, but their stress at break is reduced. The incorporation of TPU/APP systems leads also to an increase of the maximum elongation values, and, concurrently, to a decrease of the force necessary to break the material. Data reported in Table 1 shows that different mechanical performances may be obtained by the selection of the TPU.

Bugajny et al. (19, 20) have recently used the S90-based formulation as a "probe" material for the evaluation of the dynamic properties of an intumescent char. They showed that a limit in the intumescence development exists which coincides with the loss of the protective character of the coating. The formation of the intumescent shield leads to a significant increase of the viscosity values and is shown by a peak on the curves of the apparent viscosity versus temperature and time. This implies several changes in the material viscoelastic behavior: the significant increase of the elastic part in the heat-treated material explains the physical resistance to strain and then the degradation of the intumescent protection.

Table 1: Young modulus, break point and maximum elongation of the materials.

Additive	Young modulus (MPa)	break point (MPa)	elongation at break (%)
PP	1.76	33.5	8
PP/APP	1.90	26.0	5
A	0.62	20.0	42
B	0.49	17.5	14
C	0.63	20.0	58
S90	0.84	21.5	12
S85	0.67	22.0	42
S74	0.79	23.0	4

Polyamide-based additives

FR performance

In terms of LOI a synergistic effect is observed in both EVAx-APP/PA-6 and EVAx-APP/PA-6-nano formulations, x = 8, 14 or 24 (a typical example is presented in Figure 4). This effect is observed at APP/PA-6 mass ratios equaling 3 (*9, 21-22*). One can observe that the use of PA-6-nano improves the LOI from 32 vol.-% without exfoliated clay to 37 (±1) vol.-% with clay at APP/PA-6 = 3 (wt/wt). A V-0 rating is achieved for 13.5 ≤APP≤ 34 wt.% without clay and for 10 ≤APP≤ 34 wt.% with clay (the total loadings in APP/PA-6 and APP/PA-6-nano remain equal at 40 wt.-%).

The RHR values of the flame retarded polymers are strongly reduced in comparison with the virgin EVA-24 (Figure 5). The use of PA-6nano also improves the FR performance : RHR peak = 320 (±30) kW/m^2 with PA-6 and RHR peak = 240 (±25) kW/m^2 with PA-6-nano (7, *9, 21-22*).

The FR synergistic effect of polyamide/APP additive systems have been observed in several thermoplastic matrixes. In most of these, the use of an interfacial agent is needed to obtain compatible blends.

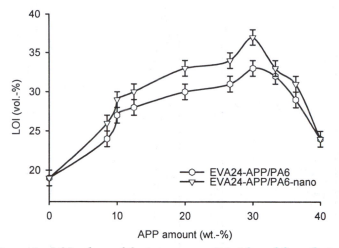

Figure 4 : LOI values of the intumescent PA-6-based formulations versus the APP content (Reproduced with the permission from reference 6. Copyright 2000 BCC)

Our laboratory has developed a silica gel coating of APP using a tetraethoxysilane (TEOS) (or a tetramethoxysilane (TMOS)) hydrolysis process (*23*) and compatible polymer/APP/PA materials via blending with addition of a functionalized polymer (EVAx copolymers (6, 9), ethylene-methyl acrylate copolymers (24), ethylene-butyl acrylate copolymers (25), ethylene-butyl acrylate-maleic anhydride copolymers (EBuAMA) (26), ethylene -maleic anhydride copolymers or propylene- maleic anhydride copolymers (24)). The effect of the constituent monomers in ethylenic co- and terpolymers has been studied earlier (25).

Typical FR blends and their performances are reported in table 2. These blends may be used in electrical (low tension cables and wires, battery shells), housing (floor) and household electricals applications. Other polyamide grades (PA6,6, PA4,6, PA11 and PA12) have also been tested as charring agent. No application can be proposed because of processing problems (high melting temperature or polymer degradation during processing) and/or poor FR performances

Mechanical performance

Figure 6 shows the mechanical properties of intumescent EVA24-based materials in comparison with the virgin polymer and a classical FR Aluminium trihydrate (ATH)-based formulation. Among the FR polymers, stress and elongation at break are the highest for the formulation containing PA-6nano.

Table 2: Composition and FR performances of intumescent blends.

Polymer matrix	PA-6/APP kg/kg	Additive loading kg.-%	Interfacial agent	LOI Vol.-%	UL94 rating
EPR	1/3	35	EBuAMA	24	no rating
EPR	1/3	35	EVA5	25	no rating
EPR	1/2	35	EBuAMA + TEOS	26	no rating
EPR	1/2	35	EBuAMA + TMOS	27	no rating
PP	1/3	35	EBuAMA	30	V0
PS	1/3	30	EBuAMA	26	no rating
HIPS	1/3	45	EBuAMA	30	no rating
HIPS	1/2	45	EVA8	33	no rating
HIPS	1/2	50	none (blooming)	42	V0
EVA8	1/5	30	TEOS	31	V0
EVA8	1/5	40	TEOS	35	V0
EVA8	APP alone	40	TEOS	30	V0
PP+talc (20 wt.-%)	1/2 or 1/3	45	EBuAMA	30	V0

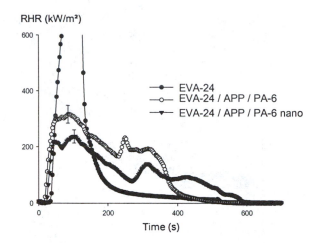

Figure 5 : RHR values versus time of virgin EVA24, EVA24-APP/PA-6 and EVA24-APP/PA-6-nano (heat flux = 50 kW/m²) (Reproduced with the permission from reference 6. Copyright 2000 BCC).

These results show the benefits of using PA-6 and PA-6 clay hybrid as charring polymers in intumescent formulations. Moreover, the use of PA-6 improves both mechanical and fire properties of FR EVA-based materials. Bourbigot et al. (7-8) proposed that the nano-dispersed clay allows the thermal stabilization of a phosphorocarbonaceous structure in the intumescent char, which increases the efficiency of the shield and, in addition, the formation of a "ceramic" which can act as a protective barrier and thus limit the oxygen diffusion towards the polymer.

Influence of aging on FR and mechanical performancesof the intumescent blends

In this last section, modifications of the performances of the EVA24/APP/PA6 blend by an artificial weathering are reported. Other recent studies (9) show that migration and hydrolysis of APP do not result from light and weathering, water absorption (ASTM 570-ISO 62 and DIN 53495) or long term immersion (10 days according to NF T 51-166).

Figure 7 shows that the light and weathering test leads to a decrease of the ignition time which may be explained by a degradation of the polymeric matrix via a free radical process, leading to the formation of conjugated polyenes in the EVA chain with a partial breaking of its chains. The formation of the unsaturated species is verified by the yellowing of the sample ($\Delta L < 0$, $\Delta a > 0$ and $\Delta B > 0$ in the Lab color space (NF EN 20105-A02 – December 1994).

146

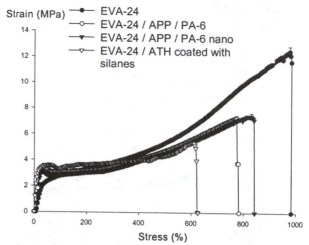

Figure 6 : mechanical properties of virgin EVA24, EVA24-ATH,
EVA24-APP/PA-6 and EVA24-APP/PA-6-nano.

**Table 3: comparison of the traction performances of EVA24, "fresh"
EVA24/APP/PA6 and aged EVA24/APP/PA6 formulations.**

Samples	Strain at break (MPa)	Elongation at break (m-%)
EVA24	12.1	982
"Fresh" EVA/APP/PA6	7.3	710
Aged EVA/APP/PA6	4.6	520

The interesting results of this study consist in the comparatively low values of the heat released (figure 7), the volume of smoke production (figure 8) and the carbon monoxide production from the aged sample. They show that the aging process (photochemical aging due to structural defect and oxidation with formation of peroxyl radical) increases the selectivity for charring of the thermo-oxidative degradation of the intumescent material.

The observed fire behavior agrees well with our previous proposition, i.e. the intumescent material degradation (surface ablative process) has to occur quickly (low ignition time) to insure the protection of the residual polymer (1, 3 25) and the material has to present a large amount of free radicals (initial peroxyl species) to trap the hydrocarbon chains evolved from the polymeric matrix degradation (26, 27).

Data in Table 3 show a loss of the mechanical properties of the intumescent material after the light and weathering test. Nevertheless, a comparison with the

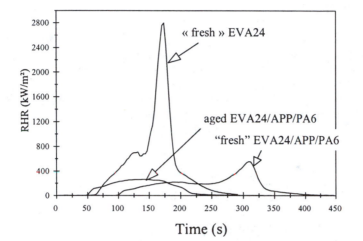

*Figure 7 : rates of heating rate of "fresh" EVA24, "fresh" EVA24/APP/PA6
and of aged EVA24/APP/PA6 formulations (external heat flux: 50 kW/m²).*

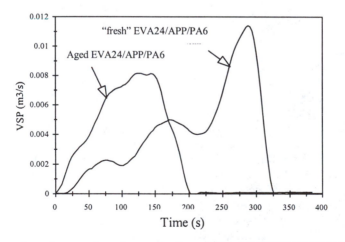

*Figure 8 : volumes of smoke production of "fresh" EVA24/APP/PA6 and of
aged EVA24/APP/PA6 formulations (external heat flux: 50 kW/m²).*

elongation behaviors of classical FR ATH-based EVA24 formulations (with and without addition of zinc borate (BZn) (*28*)) shows that the performance of the aged formulation remains comparatively high (*9*).

Conclusion

The paper present two interesting families of intumescent blends. It reports testing for miscibility of the polymer phases and sizing of the ammonium polyphosphate particles, some of them allowing a synergistic effect. It is also shown that blending allows one to obtain materials with both interesting FR performances and very different mechanical properties.

References

1. Delobel, R., Le Bras, M., Ouassou, N., Alistiqsa, F., *J. Fire Sci.* **1990**, *8*, 85-109.
2. Le Bras, M., Bourbigot, S., Le Tallec, Y., Laureyns, J., Polym. Deg. Stab. **1997**, 56, 11-21.
3. Le Tallec, Y., Ph.D. thesis, USTL, Lille, France, 1992.
4. Le Bras, M., Bourbigot, S. In *Polypropylene- An A-Z Reference*; Karger-Kocsis, J, Ed.; Kluwer Academic Publishers: London, UK, 1999; pp. 357-365.
5. Bugajny, M., Le Bras, M., Bourbigot, S., Poutch, F., Lefebvre, J.-M.., *J. Fire Sci.* **1999**, *17*, 494-513.
6. Le Bras, M., Bourbigot, S., Félix, E., Pouille, F., Siat, C., Traisnel, M., *Polymer* **2000**, *41*, 5283-5296.
7. Bourbigot, S., Le Bras, M., Dabrowski, F., Gilman, J.W., Kashiwagi, T., *Fire & Mater.* **2000**, *24, 201-208.*
8. Bourbigot, S., Le Bras, M., Dabrowski, F., Gilman, J.W., Kashiwagi, T. In *Recent Advances in FR of Polymeric Material*, Lewin, M., Ed., Volume 10; Business Communications Co Inc.: Norwalk, CT, 1999, pp. 100-120.
9. Siat, C., Ph.D. thesis, Université d'Artois, Lens, France, 2000.
10. Standard test method for measuring the minimum oxygen concentration to support candle-like combustion of plastics, ASTM D2863/77.: American Society for Testings and Materials: Philadelphia, PA, 1977.
11. Tests for flammability of plastics materials for part devices and appliances, Underwriters Laboratories, ANSI//ASTM D-635/77: Northbrook, IL,1977.
12. Babrauskas, V., *Fire & Mater.* **1984**, *8*, 81-95.

13. ASTM D 2565-ISO 4892-DIN 53387.
14. *International Plastics Handbook*, Wooebken, W, Haim, J, Hyatt, D., Eds.; 3rd Edition, Hanser Publishers, Munich, G, 1995, p.449.
15. Bourbigot, S., Le Bras, M., Delobel, R., Bréant, P., Trémillon, J-M, *Polym. Deg. Stab.* **1996**, *53*, 275-287.
16. Bugajny, M., Le Bras, M., Bourbigot, S., Delobel, R., *Polym. Deg. Stab.* **1999**, *64*, 157-163.
17. Bugajny, M., Le Bras, M., Bourbigot, S., *J. Fire Sci.* **2000**, *18*, 7-27.
18. Bourbigot, S., Delobel, R, Le Bras, M., Schmidt, Y., *J. Chim. Phys.* **1992**, *89*, 1835-1852.
19. Bugajny, M., Le Bras, M., Noël, M., Bourbigot, S., *J. Fire Sci.* **2000**, *18*, 104-129.
20. Le Bras, M., Bugajny, M., Lefebvre, J-M., Bourbigot, S., *Polym. Int.* **2000**, *49*, 1-10.
21. Siat, C, Le Bras, M, Bourbigot, S, *Fire & Mater.* **1998**, *22*, 114-128.
22. Le Bras, M., Bourbigot, S., Revel, B., *J. Mater. Sci.* **1999**, *34*, 5777-5782.
23. Le Bras, M., Bourbigot, S., unpublished results; In INPI plis Soleau 59305 (May 30th 1996) and 16414 (June 8th 1998).
24. Bourbigot, S., Le Bras, M., Delobel, R., Trémillon, J.-M., *J Chem. Soc. Faraday Trans.*, **1996**, *92*, 3435-3444.
25. Bourbigot, S., Le Bras, M., Delobel, R., Bréant, P., Trémillon, J.-M., *Polym. Deg. Stab.* **1996**, *54*, 275-287.
26. Bourbigot, S., Bréant, P., M., Delobel, R., Nathiez, P, European Patent 94401317.6, to Elf Atochem SA, 1994.
27. Le Bras, M., Bourbigot, S. In *"Fire Retardancy of Polymers - The Use of Intumescence"*, Le Bras, M., Camino, G., Bourbigot, S., Delobel, R., Eds., The Royal Society of Chemistry, Cambridge, UK, 1998, pp.64-75.
28. Le Bras, M, Bourbigot, S, Delporte, C, Siat, C, Le Tallec, Y., *Fire & Mater.* **1996**, *20*, 191-203.
29. Bourbigot, S., Le Bras, M., Leeuwendal, R., Shen, K.K., Schubert, D., *Polym. Deg. Stab.* **1999**, *64*, 419-425.

Chapter 12

The Effect of Aluminosilicates on the Intumescent Ammonium Polyphosphate–Pentaerythritol Flame Retardant Systems: An X-ray Photoelectron Spectroscopy Study

Jianqi Wang, Ping Wei, and Jianwei Hao

National Laboratory of Flame Retardant Materials, School of Chemical Engineering and Materials Science, Beijing Institute of Technology, 100081 Beijing, China

The influence of zeolite and clay (montmorillonite) on thermal degradation of the classical ammonium polyphosphate/ pentaerythritol (APP/PER) system was explored by X-ray Photoelectron Spectroscopy (XPS), a surface sensitive technique. These aluminosilicates do modify the degradation reaction which occurs between APP and PER. Information on the ratios of O/C, P/N and Si/Al are calculated and it was found that some of these change with temperature. This data in then used to understand the degradation process.

Dedicated to Professor Edgar Heilbronner on the occasion of his 80th birthday.

Introduction

The interaction between ammonium polyphosphate (APP) and pentaerythritol (PER) is of primary importance in intumescent flame retardant (IFR) systems, which have been extensively studied *(1)*. Bourbigot and Le Bras carried out an extensive study on the effect of chemical and structural aspects of zeolites and clays on the flammability performance of the APP/PER intumescent FR system. When the mass fraction of aluminosilicates is 1.5 %, they reported that the zeolitic structure, rather than the clay, leads to the best FR performance *(2,3)*. They have also studied these systems by X-ray Photoelectron Spectroscopy (XPS) *(4, 5)* and they provide a significant amount of information on the influence of the aluminosilicate structure on degradation and charring in the APP/PER system, through C1s, O1s, N1s, and P2p windows.

This study may be viewed as an extension of that work in which the surface concentrations of the elements, including Al and Si, are determined and the ratios of these various elements are used to provide information on the course of the reaction.

Experimental

Materials

The materials used in this study were all acquired from commercial sources: pentaerythritol (PER), Beijing Tongxian Chemical Engineering Factory; Ammonium polyphosphate (APP), Hoechst AG; zeolites 4A (NaA, Si/Al = 1.0) and 13X (NaX, Si/Al = 1.23), Dalian Catalyst Factory; Montmorillonite (Si/Al = 2.0), Southern Clay products, Inc.; SiO_2, Tianjin Carbon Black Factory; and Al_2O_3, Beijing Institute of Mining and Metallurgy.

Samples were prepared by grinding a 3:1 mixture by mass of APP and PER for 4 hrs, followed by combining this mixture with the aluminosilicate in a 14:1, aluminosilicate:APP+PER, mass ratio and grinding for an additional 4 hrs.

XPS (X-ray Photoelectron Spectroscopy) experiments were recorded on Perkin-Elmer Φ5300 ESCA system (Mg K_α) at 250 W under a vacuum better than 10^{-6} Pa. The instrument is equipped with a differential pumping system and a position sensitive detector (PSD) and runs under computer control. Both qualitative and quantitative XPS data was acquired from carbon (C1s), phosphorus (P2p), nitrogen (N1s), oxygen (O1s), silicon (Si2p), aluminum (Al2p) spectra as a function of temperature. Samples with thickness in micron range were carefully prepared and cast on gold foil, rather than the aluminum foil normally used for these studies, in order to enable the observation of the aluminum signals from samples. The pseudo-in-situ XPS approach that has been commonly used in this laboratory was employed*(6-9)*.

Results and Discussion

XPS Study of the Thermal Degradation of APP/PER/SiO$_2$, APP/PER/Al$_2$O$_3$ and Aluminosilicates

The conventional technique for mechanistic studies of thermal degradation of polymers is to analyze the volatile products which are evolved upon heating. In previous work from this laboratory (10-11) and other laboratories (4,5), the utility of XPS to evaluate surface chemistry has been shown. In this work, the ratios of oxygen to carbon and of phosphorus to nitrogen and particularly silicon to aluminum during the thermal degradation of APP/PER in the presence of both zeolites, 4A and 13X, and montmorillonite were monitored by XPS. It must be remembered that XPS is much more sensitive than thermogravimetric analysis (TGA) so the temperatures will always be lower in an XPS study.

The O/C ratio

The published TGA data on PEDP (pentaerythritol diphosphate), a model compound for the mixture of APP and PER, by Camino, et al. (12, 13) reveals that the degradation occurs in three steps: 100°-340°C, 340°-410°C, > 410°C. Ammonia and water are evolved during the first step; the maximum intumescence occurs between 280° and 350°C and water is the major volatile product evolved. A number of volatile organics, e.g. methane, ethylene, propylene, acetaldehyde and higher unsaturated aldehydes (C < 5) were also found on heating at 285°C (14, 15, 16). Thus the O/C ratio should decrease with temperature initially as water is lost but may increase at higher temperatures as organics are evolved. Pyrolysis of organic phosphates may also volatilize olefins and this will also effect the ratio (13,14). Figure 1 shows the O/C ratio as a function of temperature for APP/PER alone and in the presence of silica and alumina. The major differences in these curves is the position of the minima; when no additional material is present, the minimum in the O/C ratio occurs at 330°C and this is lower in the presence of silica (260°C) and alumina (270°C). This suggests that the presence of these materials permits an easier degradation.

The P/N ratio

Pure APP (17) begins to degrade just after melting (220°-280°C) and the evolution of ammonia and water occurs in the temperature range between 300° and 420°C. This must cause the P/N ratio to significantly increase. The results are shown in Figure 2; above about 150°C the ratio always exceeds the 1:1 value found in APP. Two possibilities can be suggested to explain the observations for the degradation of APP/PER alone. Between 100° and 350°C the lack of nitrogen atoms at the surface must result from the evolution of ammonia, because the mass loss of phosphorus can be neglected. However, above 350°C the dramatic increase in P/N may arise from the elimination of ammonia and the subsequent cross-linking, for example, the "ultraphosphate" structure proposed by Camino, et al., *(17)*. The effect of the extent of cross-linking on the signal intensity in XPS, i.e. the number of P atoms per unit area, has been discussed previously in the literature *(9, 18)*. The second possibility may be ascribed to degradation of the phosphate ester and migration of the degraded fragments, e.g. PO· and PO·$_2$., where they can react with the degrading surface. Finally it should be noted that SiO_2 has a much greater effect on the ratio than does Al_2O_3, perhaps indicative of some catalytic effect.

Evidence for the dynamic motion of components in the surface may be found by determining the atomic concentration of Si and Al at the surface. Both vary with temperature, as seen in figure 3. There is a dramatic decrease in silicon and a small increase in aluminum concentration, showing that these atoms are quite mobile and can migrate to or away from the surface. The temperature-dependent dynamics of outward and inward movement depends on the interfacial energy between the phases surrounded by the environment of the swollen char; this topic merits further investigation.

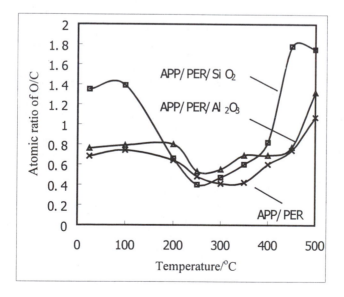

Figure 1. Dependence of the O/C ratio for APP/PER, APP, PER/SiO_2, and APP/PER/Al_2O_3 with temperature.

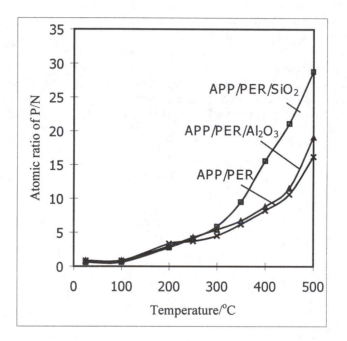

Figure 2. Dependence of the P/N ratio for systems of APP/PER/SiO$_2$ and APP/PER/Al$_2$O$_3$ on temperature

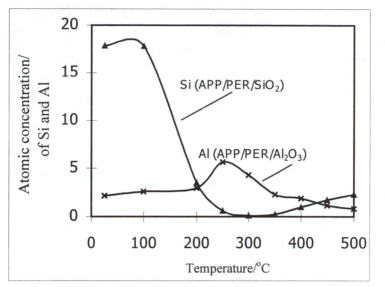

Figure 3. Atomic concentration of Si and Al for systems of APP/PER/SiO₂
and APP/PER/Al₂O₃ on temperature

Aluminosilicate systems

Systems of APP/PER/Aluminosilicates can be investigated in a similar way. For reference, some of our work on the thermal behavior of APP/PER systems in the presence of aluminosilicates, such as zeolites 4A, 13X, mordenite, ZSM-5 and montmorillonite) by TGA experiments *(19)* is introduced here. The DTG curves for zeolites 4A, 13X and montmorillonite with APP/PER show similar trends: the DTG peak is at lower temperature, with zeolite 13X causing the smallest shift, followed by montmorillonite and zeolite 4A shows the largest shift while the mass loss rate is lowered in the presence of all of these materials, with zeolite 13X giving the smallest mass loss rate, followed by zeolite 4A and montmorillonite giving the largest mass loss rate, which is still one-half of that of the simple blend of APP and PER.

The O/C ratio

The data are collected in figure 4. The curve for montmorillonite lies above the others, implying that the surface of clay was heavily contaminated by oxygen particularly at temperature below say 100°C. The decrease in the O/C ratio must be attributed to the evolution of water from the esterification reaction between

APP and PER. The minima for the three curves stays constant at about 250°C. Within 250°-400°C the O/C ratio steadily grows, possibly due to the unavoidable superficial oxidation and the degradation of the carbonaceous residue with the evolution of carbon-containing species. The cross-linked network formed during charring is likely to account for the decrease in the O/C ratio at temperatures above 400°C *(9, 18)*.

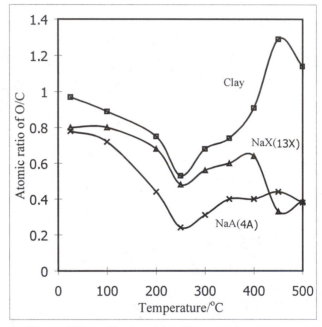

Figure 4 Dependence of the O/C ratio for systems of APP/PER/Alumino-silicates on temperature

The P/N ratio

The results on the P/N ratios for the three inorganic additives are depicted in figure 5. The P/N ratio begins to increase at 100°C and shows a rapid increase at about 250°C. The maxima occur at 300, 400 and 430°C for systems containing 13X, 4A and montmorillonite, respectively. Similar to the analysis as mentioned above for systems of APP/PER/SiO_2 and APP/PER/Al_2O_3, 13X seems to be the most effective in accelerating the release of ammonia. The pyrolytic degradation of the carbon-based phosphorous ester or some similar species, such as the "ultraphosphate," may also be responsible for the decrease in the P/N ratio at higher temperatures, perhaps due to the formation of nitrogen-containing aromatics and the evolution of phosphorus oxides. It is apparent that in the presence of aluminosilicates the systems of APP/PER/aluminosilicates do cause a modification of the degradation process, resulting in a reduction of mass loss rate in the TGA. It should be noted that over the entire range of temperature the ratio of P/N is always larger than the horizontal line, *i.e.*, the surface is always deficient in nitrogen and has an excess of phosphorus; in agreement with the literature *(4, 5)*.

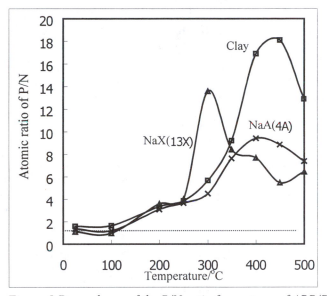

Figure 5 Dependence of the P/N ratio for systems of APP/PER/ Aluminosilicates on temperature (horizontal line represents P/N=1 for APP)

The Al/Si ratio in APP/PER/Aluminosilicates.

Zeolites and/or clays have been considered as solid acids and said to be heterogeneous catalysts for many organic reactions *(20, 21)*. Nevertheless, in flammability research, zeolites and/or clays are often used as additives at a high loading level. When aluminosilicates together with organics are subjected to heat, they do undergo chemical changes *(22, 23)*. This can also be shown from binding energies. The binding energy of Si2p and Al 2p in sodium zeolite, for example, are 101.4 eV and 73.5 eV, respectively. At 500°C, these binding energies are now found to be 103.2 eV and 74.4 eV, which are the values of SiO_2 and Al_2O_3. The conversion of zeolites and montmorillonite combined with APP/PER into other species occurs at temperature $\geq$ 200°C *(19)*. It is clear that

158

"catalyst" is not the correct term to use in this area, since it does undergo significant change. Nonetheless, the terminology catalyst is still used in many applications, for example, pyrolysis of polymers *(24)* at higher temperatures. XPS data on the Si/Al ratio in systems of APP/PER/ aluminosilicates was conducted as function of temperature. The ratio of Si/Al was determined using Si2p and Al2p windows, figure 6. Minima are observed for each system at 260° for zeolite 4A, 310° for zeolite 13X, and 350°C for montmorillonite. There is a significant rise in the ratio after this minina. On the basis of the binding energies, the zeolites and clays maintain their integrity up to at least 200° even to 250°C. At higher temperatures, the aluminosilicates undergo degradation and the differential migration of the two components to and away from the surface controls the ratio.

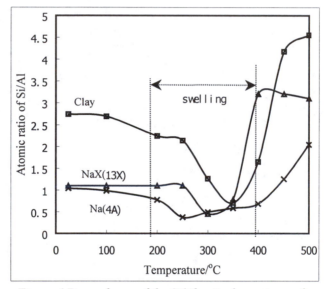

Figure 6 Dependence of the Si/Al ratio for systems of
APP/PER/Aluminosilicates on temperature

Surface aluminum phosphates, or other materials containing these elements, and/or aluminum oxide, can play an important role in the intumescent coating. These ceramic-like materials can act as an insulator in the 280-400°C temperature region where the protective effect of the intumescent coating is maximum *(25)*.

Conclusion

XPS is a useful technique to analyze intumescent fire retardant systems, since it can provide information on what is occurring at the surface of a degrading polymer. Zeolites and clays, as well as alumina and silica, can catalyze the esterification reaction between APP and PER. Both zeolites and clays undergo degradation reactions above 200°-250°C which can have an effect on the degradation pathway.

References

1. Camino, G.; Costa, L., Trossarelli, L. *Polym. Degrd. Stab.*, **1984**, 7, 25-31.
2. Bourbigot, S.; LeBras, M.; Delobel, R.; Bréant, P.; Trémillon, J. M. *Polym. Degrad. Stab.*, **1996**, 54, 275-287.
3. Bourbigot, S.; LeBras, M.; Delobel, R.; Revel, B.; Trémillon, J. M. In *Recent Advances in FR of Polymeric Materials*; Levin, M., Ed.; BCC, Stamford, Connecticut, USA, **1996**.
4. Bourbigot, S.; Le Bras, M.; Gengembre, L.; Delobel, R. *Appl. Surf. Sci.*, **1994**, 81, 299-307.
5. Bourbigot, S.; Le Bras, M.; Delobel, R.; Gengembre, L. *Appl. Surf. Sci.*, **1997**, 120, 15-29.
6. Li, B.; Wang, J. *J. Fire Sci.*, **1997**, 15, September/October 341-357.
7. Wang, J. In *Fire Retardancy of Polymers, The Use of Intumescence;* Le Bras, M.; Camino, G.; Bourbigot, S.; Delobel, R., Eds.; The Royal Society of Chemistry, Cambridge, UK, 1998; pp 159-172.
8. Han, S.; Wang, J. *Chinese Science Bulletin;* **2000**, in press.
9. Hao, J.; Wu, S.; Wilkie, C. A.; Wang, J. *Polym. Degrad. Stab.*, **1999**, 66, 81-86.
10. Wang, J.; Lang, H. *Science in China*, **1988**, B 5, 463-472.
11. Wang, J.; Feng, D.; Tu, H. *Polym. Degrad. Stab.* **1993**, 43, 93-99.
12. Camino, G.; Costa, L.; Trossarelli, L. *Polym. Prepr.* **1984**, 25, 90.
13. Camino, G.; Martinasso, G.; Costa, L. *Polym. Degrad. Stab.* **1990**, 27, 285-296.
14. Camino, G.; Martinasso, G.; Costa, L.; Gobetto, R. *Polym. Degrad. Stab.* **1990**, 28, 17-38.
15. Camino, G.; Delobel, R. In *Fire Retardancy of Polymeric Materials*; Grand, A.F.; Wilkie, C.A., Ed., Marcel Dekker, New York, **2000**; pp 217-243.
16. Camino, G.; Costa, L.; Trossarelli, L. *Polym. Degrad. Stab.* **1984**, 7, 25-31.
17. Camino, G.; Costa, L.; Luda, M. P. *Makromol. Chem. Makromol. Symp.* **1993**, 74, 71-83.
18. Hao, J.; Wilkie, C.A.; Wang, J. *Polym. Degrad. Stab.*, **2000**, in press.

160

19. Wei, P.; Wang, J. *Polym. Mat. Sci. Eng.*, **2000,** submitted.

20. Barthomeuf, D. In *Catalysts by Zeolites*; Imelik, B.; et al. (Eds.); Elsevier, Amsterdam; **1980**, pp 55.

21. Tanabe, K.; Misono, M.; Ono, Y.; Hattori, H. *New Solid Acids and Bases, Their Catalytic Properties*, Elsevier, Amsterdam; 1989, pp 128-141, 142-163.

22. Bourbigot, S.; Le Bras, M.; Delobel, R.; Trémillon, J-M. *J. Chem. Soc., Faraday Trans.*, **1996**, 92, 3435-3444.

23. Bourbigot, S.; Le Bras, M.;Bréant, P.; Trémillon, J-M.; Delobel, R. *Fire and Mat.;* 1996, 20, 145-154.

24. Zhao, W.; Hasegawa, S.; Fujita, J.; Yoshii, F.; Sasaki, T.; Makuuchi, K.; Sun J.;Nishimoto, S.I. *Polym. Degrad. Stab.*; **1996**, 53, 129-135.

25. Delobel, R.; Le Bras, M.; Ouassou, N.; Alistiqsa, F. *J. Fire Sci.* **1991**, *8*, 85-109.

Chapter 13

Role of Interface Modification in Flame-Retarded Multiphase Polyolefin Systems

Gy. Marosi, P. Anna, Gy. Bertalan, Sz. Szabó, I. Ravadits, and J. Papp

Organic Chemical Technology Department, Budapest University of Technology and Economics, H-1111 Budapest, Muegyetem rkp. 3, Hungary
(email: Marosi.oct@chem.bme.hu)

Different methods of interface modification were investigated in flame retarded polyolefin systems including non-reactive and reactive surfactants, zinc hydroxystannate and elastomers. In most cases the reactive modifications resulted in better performance.

The importance of interface modification in filled, reinforced polymers has been well discussed (1). It was found that a multilayer interphase of surfactant and elastomer layers is a good way to avoid the negative effects of inclusions (2). The polarity, viscosity and melting regime affect the fraction of elastomer in the interphase and disperse phase and thus the structure-property relationship (3). Reactive surfactants were developed in order to combine the function of dispersing and coupling agents (4).

A large number of flame retardant additives are applied in the form of particles dispersed in the polymer matrix (5). The question may arise whether the rules affecting the properties are similar to other filled systems or are there some special features for flame-retardants? To answer this question, different interfacial structures were formed in polyethylene-metal hydroxide (MOH) and polypropylene-ammonium polyphosphate (APP) systems. The principles applied to filled polymers for modification of the interfacial layer served as the basis for the modification of flame retardant (FR) systems as well.

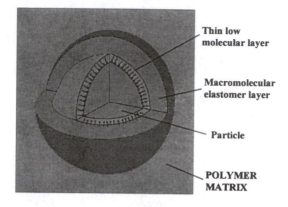

Figure 1. Scheme of multilayer interphase

A multilayer interphase, consisting of a thin and a thicker sub-layers, according to the general scheme in Figure 1, was formed. These layers, formed using reactive or non-reactive additives, were separately analyzed.

Experimental

Materials:

Polypropylene (PP): Tipplen H 337, product of Tisza Chemical Works (Hungary), propylene homopolymer; density, 0.9 g/cm^3; melt index; 12 g/10 min (21.6 N, 230°C); Tipplen K793, product of Tisza Chemical Works (Hungary), ethylene-propylene copolymer; density, 0.9 g/cm^3; melt index; 0.7 g/10 min (21.6 N, 230°C); *High density polyethylene (HDPE):* Tipelin PS 380-09, product of Tisza Chemical Works (Hungary), *Ethylene vinylacetate copolymer (EVAc):* EVA 3325 (ICI); VAc content, 33%; density, 0.98 g/cm^3; melt index, 28.8 g/10 min (21.6 N, 190°C). *Ammonium polyphosphate (APP):* Exolit 422, $[NH_4(PO_3)]_n$ (n=700); Clariant (Germany). *Melamine polyphosphate (MPP):* Melapur 200, commercial sample of DSM. *Al(OH)$_3$ (AH):* Alolt 50AF (AH I.); BET area, 3 m^2/g; Alolt 59 (AH II.), BET area, 4 m^2/g; ground types, Alolt 60FLS (AH III.) precipitated type, BET area, 6 m^2/g; Ajka Alumina Co. Ltd. (Hungary). *Mg(OH)$_2$, (MH):* Magnifin H5; Magnesiaprodukte GmbH (Austria), *Zinc hydroxystannate coated Mg(OH)$_2$ (MH/ZHS):* experimental product of International Tin Research Institute Ltd. (Uxbridge, UK). *Non-reactive* *surfactant (S)* was Estol 1474 type glycerol monostearate (GMS), Mp., 60°C (Unichema International). *Reactive surfactant (RS)* was a derivative of glycerol monostearate, a polycondensable (pre-condensed) polyol. The chemical structure and preparation of RS and *Silicone additive (Sil)* are described elsewhere (*6*).

Methods:

   The preparation of additives was carried out using a computer-controlled reactor constructed in our laboratory in order to develop controlled organic chemical processes which measure their heat effects and viscosity changes in line. The hardware and software descriptions of the reactor calorimeter system are given elsewhere (*7*). *Compounding*, for forming the flame retardant polymer systems, was performed using a Brabender Plasti-Corder PL2000 with a rotor speed 50rpm at 200°C, while a laboratory press was used for *forming compression-molded sheets* of 1.6 mm thickness. *Differential Scanning Calorimetry (DSC)* measurements were performed using Setaram DSC 92, sample weight: 10 mg, heating rate: 10°C/min, atmosphere: air. The samples were heated to 50°C higher than their melting temperatures in order to remove their thermal history (melting memory effect) (*8*) and then the crystallization and heating curves were determined. *Thermal Gravimetry (TG)* measurements were performed using Setaram Labsys TG equipment, sample weight: 50 mg, heating rate: 20°C/min, atmosphere: N_2. *X-ray photoelectron spectroscopic (XPS)* characterization of samples was performed by Kratos XSAM 800 spectrometer using Mg $K\alpha_{1,2}$ radiation. The spectra were referenced to the hydrocarbon type carbon at binding energy, BE=285.0 eV. *Limited Oxygen Index* (LOI) was determined according to ASTM D 2863.

Results and Discussion

   The modification of the interfacial layers was performed in some cases through pretreatment of additives; in most cases the in line methods were preferred in which the polarity, viscosity and melting regime of interfacial additives were selected to promote the formation of the required structure (*3*).

Interfaces around metal hydroxides

   Metal hydroxides offer an environmental solution for flammability problems. However, a high loading (55-65%) is required for acceptable efficiency (*9,10*) and the deterioration of mechanical properties at high filler content is a limiting factor of the application. The efficiency of these additives is also affected by their particle size. Generally the smaller the particle size, the higher the efficiency, as long as the homogeneous distribution of particles is ensured. However, the use of smaller particles (especially below a few micrometer average particle size) limits the processability.

Both limits, the concentration and the particle size, are connected with the interface. The inorganic particles, due to their high surface free energy, form agglomerates surrounding a rigid layer of immobilized polymer(11). The higher surface area of the smaller particles makes this immobilized area more pronounced so that a physical network can be formed and this makes the system too rigid and barely processable.

The main purpose for forming an interlayer of enhanced characteristics between the flame retardant and polymer phase is to achieve a homogeneous disperse system. The possible effect of this modification on the efficiency of flame retardants is not yet clarified. We expect the reactivity of the additives to play a key role in this respect.

A thin interfacial layer formed by a low molecular surfactant is effective to overcome the particle size limit. The PE sample containing small (precipitated) $Al(OH)_3$ particles was barely processable and extremely brittle, but achieved a higher level of flame retardancy than the samples containing larger particles. Considering the effect of non-reactive (S) and reactive surfactants (RS), although both improved the processability, a clear difference can be seen in Table I. The former lost its FR character, while the latter preserved it; the increase of the concentration is disadvantageous for S but advantageous for RS.

Table I. Effect of surfactants on $Al(OH)_3$, containing PE systems

Composition	Particle Size, μm	LOI vol.%	UL-94 rating
PE+ 60%AH I.	5	28	HB
PE +60%AH II.	3	29	HB
PE+60%AH III	1.4	32	V-0
PE+60%AH III +0.5%S	1.4	31	HB
PE+60%AH III +1%S	1.4	30	HB
PE+60%AH III +1%RS	1.4	31	V-1
PE+60%AH III +5%RS	1.4	31	V-0

Note: AH I. and AH II. were prepared by grinding, while AH III. by precipitating, S and RS are GMS and its derivative resp. (see Experimental)

These results can be interpreted through rheological comparison. The plastograms shown in Figure 2 start with the highest torque value of mixing. This should be decreased through the modification to obtain a high quality dispersion. The non-reactive surfactant decreases the surface free energy of the particles and allows the mixing at lower energy, thus the torque decreased. The non-reactive surfactant at the interface brakes up the physical network so the rheological change is reflected in lower fire resistance. The behavior of reactive surfactant is different, it also decreases the mixing energy, but after achieving a homogeneous dispersion, the viscosity increases with increasing temperature. This result suggests that in this case, after homogenization, a new network starts to form at higher temperatures. (A chemical network is supposed to be induced by RS.) This unusual rheological behavior offers a balance between processability and FR performance (due to better dispersion, low viscosity at the processing temperature and more stabile structure at higher temperature). Optimization of this effect may permit one to achieve a system with good processability and high FR efficiency, even if the particle size is below 1 μm.

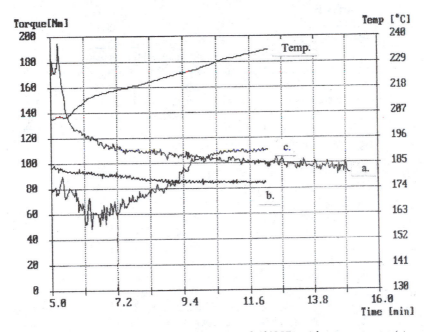

Figure 2. Plastograms of PE+60% precipitated Al(OH)₃ without treatment (a), with 1% non-reactive surfactant (b) and with 5% reactive surfactant (c)

*A **thick layer*** of elastomer interphase, replacing the rigid layer, may help to overcome the concentration limit. This limit in PE is about 60% of the FR additive, just below the effective concentration, as shown in Table II. Introducing an elastomer interlayer, formed by EVAc, the limit could be shifted above 65% and achieve acceptable flame retardancy.

Table II. Change of efficiency of Mg(OH)$_2$, due to various modifications

Composition	LOI vol.%	Tensile Str. MPa	Elong. %	Hardness Shore A
PE+60%MH	29	14	11	
PE+65%MH+EVAc*	32	12	13	
PE+60%MH+EVAc+Sil.*	42	8	113	
EVAc + 60%MH	32	4.8	109	92
EVAc + 60%MH/ZHS**	35	8	70	94
EVAc + 50%MH	24	6.2	831	
EVAc+45%MH+5%ZHS	25	8.6	606	
EVAc+50%MH/ZHS**	29	12.3	68	

Note: * EVAc and Sil. applied in an amount needed for surface modification
**Total amount of ZHS in the compound is 5%.

Zinc hydroxystannate (ZHS) interlayer around the particles may improve the flame retardancy still further. Such an interlayer has been recently found effective in PVC (*12*). According to Table II, it is also effective in EVAc and the mechanism of the improvement can be understood from Figure 3.

The larger amount of the non-extractable component (gel content), the higher torque of mixing and lower weight loss at 450°C in TG experiment are inevitable signs of catalytic cross-linking in the presence of ZHS. This means that an elastomer made reactive by ZHS coated particles is more effective than the non-reactive system. The advantage of using a reactive polymer (novolac resin) additive in MOH containing polyolefins has been recently demonstrated by Lewin (*13*).

The interlayer of reactive silicone additive resulted in the most significant improvement (see Table 2 and Figure 4). The formation of a continuous glassy layer from certain silicone additives in a flame retarded polypropylene system has been demonstrated elsewhere (*14*). The increase of the LOI rating caused by 2 % additive is probably also a consequence of the formation of a continuous glassy layer on the flame treated surface, similar to the mechanism proposed by Bourbigot (*15*) for zinc borate containing systems.

Interfaces around Polyphosphates

Polyphosphates, like ammonium polyphosphate (APP), and melamine polyphosphate (MPP), change the interfacial layer by means of a nucleating effect. This is expressed in Table III by the shift of crystallization temperature to higher temperatures. A transcrystalline interlayer surrounds the particles having nucleating activity; coating the polyphosphate particles with low molecular additive decreases this effect significantly.

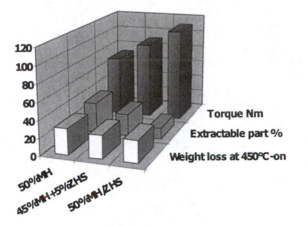

Figure 3. Effect of ZHS on torque of mixing, residue after extraction with hot hexane and TG weight loss at 450 °C of EVAc-Mg(OH)₂ system

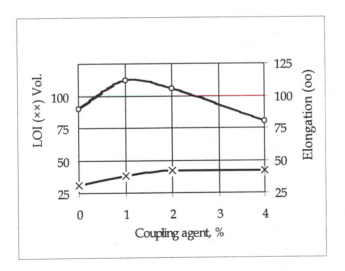

Figure 4. LOI (xxx) and elongation at break (ooo) values of 60% Mg(OH)₂ filled HDPE against the concentration of reactive silicone additive

Table III. DSC data of flame retardant PP-s

Materials	Crystallization temp. (°C)		ΔH_C, PP (J/g)	Second melting temperature (°C)		ΔH_m, PP (J/g)
	T_{C0}	T_C		T_{m0}	T_m	
Tipplen H 337	119.8	113.8	84.1	160.3	172.7	70.6
Tipplen H 337 + 18% APP	129.9	123.4	92.9	165.7	176.9	94.5
Tipplen H 337 + 18% APP coated	125.6	119.2	72.9	160.2	171.6	72.3
Tipplen K 793	118	106	-	-	-	-
Tipplen K 793 + 18% MPP	126	119	-	-	-	-
Tipplen K 793 +18% MPP coated	123	115	-	-	-	-

Note: T_{C0}, T_{m0}: temperature at the start of crystallization and melting, respectively, T_C, T_m: temperature of crystallization and melting peak

The chemisorption of the surfactant layer on APP was shown by the XPS surface analytical method (see Figure 5.). APP, after coating, was extracted with chloroform and then the quantity of the chemically bonded coating layer was determined by the decrease in P atom concentration on the surface. The formation of melamine coating layer on APP particles could be similarly evaluated by XPS measurements. The APP coated this way is characterized by outstanding hydro-thermal stability, as is demonstrated in Figure 6 by the conductivity of water used for extraction of 44 x 44 mm size compression molded samples.

The formation of an elastomer interlayer around APP particles was reported earlier (16). Several advantages, like improvement of the stability, processability and mechanical properties, were found due to the formation of this interlayer. The EVAc interphase, for example, improves the water resistance and mechanical properties, but decreases the flame retardancy of APP containing PP, therefore the use of boroxo siloxane elastomers was proposed. It may affect both the stability and FR efficiency. The improvement is in connection with the mobility of additives in the polymer systems and with the formation of an oxygen barrier protective layer (14,17). According to recently reported results, the change in rheology due to the boroxo siloxane elastomer also contributes to higher flame retardancy (18).

The distribution of the components in the FR system also has a significant effect on the result. Changing the sequence of addition of additives at compounding (with the parameters defined in the Experimental) causes a significant alteration in the LOI (see Figure 7)

Intensity / Counts *1000

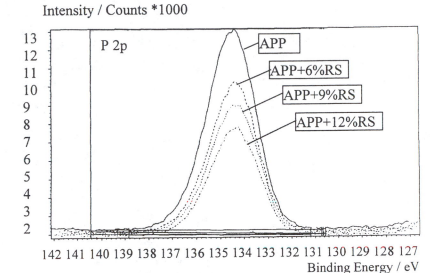

Figure 5. Amount of P atom on the surface of APP without and with surfactant coating layers (determined by XPS surface analysis)

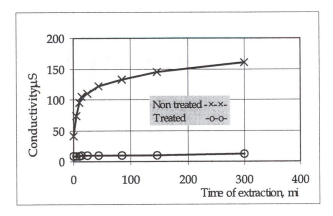

Figure 6. Conductivity of extracting water used for extraction of PP sheets containing non-treated (-x-x-) and surface treated (-o-o-) APP/polyol system

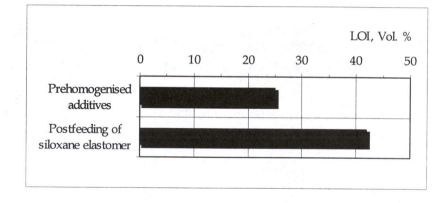

Figure 7. Influence of mixing method on FR effectiveness of intumescent FR-PP compound (mixing time 10 min)

Only the first method facilitates the formation of elastomer interlayer. According to Figure 7 the organosiloxane elastomer in a separate phase is more efficient than the interphase. In case of fire this structure promotes the quick *in situ* transformation to a continuous silicate oxygen-barrier surface layer.

Conclusions

Interfaces determine both the mechanical and stability characteristics of flame retarded polymers, even without modification.

The immobilized interphase around high concentration and low particle size FR additives of high surface free energy may form a rheologically unfavorable physical network structure. A well-planned modification is required in order to achieve better performance, which means in most cases the formation of a multilayer interphase.

A thin interlayer of low molecular surfactants improves the processability and mechanical properties by suppressing the formation of agglomerates, immobile and transcrystalline interlayers. A reactive surfactant contributes to better flame retardancy through increasing the viscosity at high temperature.

A thick interphase of elastomers allows a higher concentration of additives without deterioration of the properties, due to better stress distribution at the interfaces. An elastomer made reactive by ZHS coated particles achieves a higher level of flame retardancy than the non-reactive system.

Organosiloxanes of specially designed structure improve the rheological characteristics and may promote the transport of flame retardant additives towards the surface only in the circumstances of flame attack. In order to form an oxygen barrier layer, it is preferred to have the precursor separate from the interphase.

It can be concluded from the results that some rules affecting the properties of flame retardant polymer systems are similar to those relating to inert filler containing systems, but the picture is much more complex and there are some special features which are relevant only for flame retardants.

Acknowledgement.

The authors are grateful to Clariant and DSM for supplying the polyphosphate samples, to Dr. Cusack (International Tin Research Institute Ltd.) for preparing the ZHS coated samples, to Dr. Toth and Dr. Bertóti for their useful advises and contribution to the XPS results. The work was fulfilled with the financial support of OTKA T 026182, OTKA T 032941, and EU5 G5RD-CT projects.

Chapter 14

Thermal Degradation and Combustion Mechanism of EVA–Magnesium Hydroxide–Zinc Borate

Serge Bourbigot[1,*], Fabien Carpentier[2], and Michel Le Bras[2]

[1]Laboratoire de Génie et Matériaux Textiles (GEMTEX), UPRES EA2161, Ecole Nationale Supérieure des Arts et Industries Textiles (ENSAIT), BP 30329, 59056 Roubaix Cedex 01, France
[2]Laboratoire de Génie des Procédés d'Interactions Fluides Réactifs-Matériaux (GEPIFREM), UPRES EA2698, Ecole Nationale Supérieure de Chimie de Lille (ENSCL), BP 108, 59652 Villeneuve d'Ascq Cedex, France

The action of a zinc borate in copolymer ethylene-vinyl acetate / $Mg(OH)_2$ formulations is studied. It is shown that zinc borate acts as a synergistic agent. The mode of action of the flame retardants is elucidated. The transformation of $Mg(OH)_2$ to MgO leads in the first step to a decrease in the temperature of the substrate (endothermic effect of the decomposition of $Mg(OH)_2$); this leads to an increase in the time to ignition of the material. In a second step, a protective MgO-based ceramic is formed. It is shown that zinc borate plays the role of a binder (formation of boron oxide) in the formation of the MgO-based ceramic. This material at the surface of the substrate act then as a physical/thermal barrier. The degradation of the polymeric matrix is slowed down and the flow of flammable molecules is reduced.

Introduction

Halogen compounds are widely used for flame retarding polymers but the corrosiveness and toxicity of their combustion products and the smoke production have caused much concern. Therefore, the present approach aims to limit the use of halogen-based flame-retardant (FR) systems and turn towards halogen-free FR formulations. One approach to a solution is to add metal hydroxides, such as magnesium hydroxide, to a polymer matrix (1).

Previous studies demonstrated that there are major advantages in using a combination of zinc borates with other flame retardants in several kind of polymers (EVA, PVC, nylons…), in particular in halogen free systems (2)(3). Zinc borates and, in particular Firebrake® 415, $4ZnO.B_2O_3.H_2O$ (FB415), act as synergistic agents in FR EVA-based formulations, such as magnesium hydroxide $(Mg(OH)_2)$ filled EVA. Figure 1 shows the synergistic effect in EVAx-Mg(OH)2/FB415 formulations (EVAx, EVA with 8% or 24% (wt/wt) vinyl acetate). Moreover, UL-94 V-0 rating (ANSI / ASTM D-635-77) can be achieved.

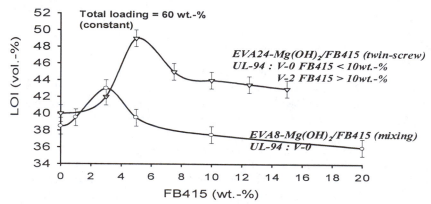

Figure 1. LOI values (ASTM D2863/77) versus the substitution of zinc borate in EVAx-Mg(OH)₂/FB415 (total loading remains constant at 60 wt.-%; processed using a Brabender mixer or a counter rotating twin screw extruder).

Oxygen consumption calorimetry (cone calorimeter) experiments (Figure 2) show that zinc borate decreases the maximum rate of heat release (RHR) and delays the second heat release peak. During these tests, the formation of a char layer at the surface of the flame retarded polymers is observed. Moreover, it is shown that the effect of FB415 addition in $EVA8/Mg(OH)_2$ formulations increases as the amount increases. In this case, no synergistic effect is observed from the LOI curves. This has been explained (4) by the fact that a high FB415 level favors the dripping process which takes place during the LOI test. When the viscosity becomes too low (FB415 > 3 wt %), the MgO-based ceramic bound by FB415 flows away from the surface of the polymer. The polymer then loses part of its protection. As a consequence, the combustion of the polymer is favored and the LOI performance becomes poor.

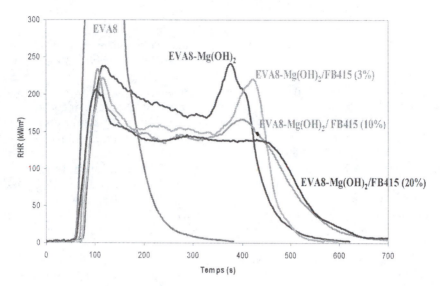

Figure 2. RHR of EVA8-Mg(OH)₂ formulations with and without zinc borate under an external heat flux of 50 kW/m² (ASTM 1356-90 and ISO 5660).

In this paper, the mechanisms of the thermal and fire degradations of the FR EVA8-based formulations are investigated using ^{13}C, ^{25}Mg and ^{11}B NMR of the solid state. The modes of action of FB415 are then discussed.

Experimental

EVA8 (Evatane) was supplied by Elf Atochem. Magnesium hydroxide (Mg(OH)$_2$) was a commercial grade (DEAD SES BROMINE MHRM 100), and FB415 was supplied by US Borax®. Table 1 gives the composition of the formulations.

Table 1. Composition of the FR formulations.

Formulation	EVA8 (wt.-%)	Mg(OH)₂ (wt.-%)	FB 415 (wt.-%)
EVA8- Mg(OH)$_2$	40	60	0
EVA8- Mg(OH)$_2$/FB415	40	57	3

Mixing was first carried out in a Brabender roller mixer measuring head (mixer E350 with roller blades, volume 370 cm^3 at 160°C and with 30 rpm rotor speed).

Thermogravimetric analyses were carried out at heating rate 7.5°C/min under synthetic air (Air Liquid grade; flow rate = 5.10^{-7} Nm^3/s) using a Setaram MTB 10-8 thermobalance. In each case, samples (10 mg) were positioned in open vitreous silica pans. The precision on the temperature measurements is ± 1.5°C in the range 50°C-850°C.

^{13}C NMR measurements were performed on a Bruker ASX100 at 25.2 MHz (2.35 T) with magic angle spinning (MAS) and high power ^{1}H decoupling (DD) using a 7 mm probe. ^{1}H-^{13}C cross polarization (CP), a contact time of 1 ms, a repetition time of 10s and a scan number of 2048 (materials) or 10000 (chars) were used. The reference used was tetramethylsilane and the spinning speed was 5000 Hz.

^{25}Mg NMR measurements were performed on a Bruker ASX400 at 24.5 MHz (9.4 T) using a 7 mm MAS probe. In order to observe the complete FID and so, to get an undistorted spectrum, Hahn spin echo (90°-τ-180°-τ-acquire) was used. A repetition time of 10 s was used for all samples and 5000 scans were used to obtain a good signal to noise ratio. The reference used was a saturated solution of MgSO$_4$ and the spinning speed was 7000 Hz. The simulation of the spectra was made using homemade software (Quasar) (5).

^{11}B NMR measurements were performed on a Bruker ASX400 at 128.3 MHz (9.4 T) using a 4 mm MAS probe. A repetition time of 10 s and 128 scans were used for all samples. The reference used was (C$_2$H$_5$)$_2$O,BF$_3$. The spinning speed was 15000 Hz for all samples. The simulation of the spectra was made using the Quasar software.

Results and Discussion

The thermo-oxidative degradation of EVA8 occurs in three steps (Figure 3) and has been already discussed elsewhere (6). The FR polymers are more thermally stable than the virgin material (Figure 3). They have a similar thermal behavior with and without FB415. Firstly, the degradation of the material corresponding to the first degradation step of EVA8 occurs at higher temperatures than the virgin EVA8 (temperature range: 300-400°C). Secondly, one can notice that FB415 slightly increases the thermal stability of the formulations in the temperature range 450-500°C.

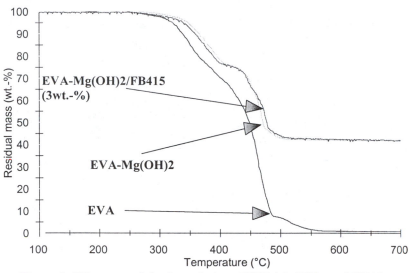

Figure 3. TG curves of the formulations EVA8-Mg(OH)₂ and EVA8-Mg(OH)₂/FB415 in comparison with the virgin EVA8.

To simulate the thermal degradation of the FR materials, they were heat treated, at four different characteristic temperatures (200, 300, 380 and 500°C) for 15 hours under air flow. MAS ^{25}Mg NMR spectra versus temperature of the two formulations show one or two species (300°C) (Figure 4 and Figure 5). They can be assigned to $Mg(OH)_2$ (isotropic chemical shift ≈ 15 ppm computed from spectrum simulation) and MgO (isotropic chemical shift ≈ 26 ppm). This evolution agrees with the classical mode of action of dehydration of $Mg(OH)_2$ to MgO proposed in the literature (7). It is interesting to note that FB415 slows the kinetics of degradation of $Mg(OH)_2$ into MgO at 300°C. Indeed the ratios $Mg(OH)_2$/MgO are respectively 0.54 and 0.81 in the case of the heat-treated formulations without and with FB415. This result suggests therefore that FB415 improves the FR performance of the formulation by "spreading" the activity of $Mg(OH)_2$. This latter would be active longer and would preserve a lower temperature in the substrate longer.

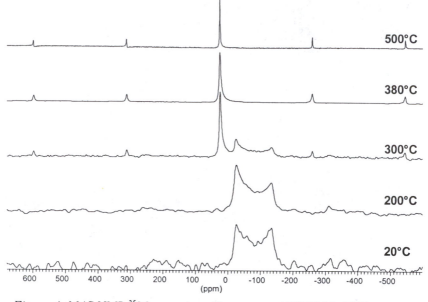

Figure 4. MAS NMR ^{25}Mg spectra of heat-treated EVA8-Mg(OH)$_2$ versus temperature

In order to investigate the flaming degradation of EVA8-Mg(OH)$_2$/FB415, the heat flux (during a cone calorimeter experiment) is shut down at different characteristic times (40s : heating of the materials; 110s : materials formed after the maximum of RHR; 270s : pseudo steady state combustion and 410s : second RHR peak), the sample is removed from the cone calorimeter and quenched in the air before NMR analysis. Before ignition and at the end of the combustion, only one phase is observed whereas at 110 and 270s two phases can be distinguished.

No difference is observed by ^{25}Mg NMR between the formulations with and without FB415 (typical example in Figure 6). Before ignition, two species are observed which can be assigned to Mg(OH)$_2$ and MgO. Mg(OH)$_2$ is then degraded during the heating of the material to form MgO. The reaction is strongly endothermic and evolves water. This means that this transformation is responsible for the increase of the time to ignition (time to ignition of the virgin EVA8 is 30s in comparison with 75s for FR EVA). At higher times, only one species assigned to MgO is observed. It can be proposed that the protection occurs via the formation of a ceramic-like structure (MgO ceramic). For the first time it is demonstrated that the mode of action of metal hydroxides arises not only from a dilution of the flamable gases and from a strong endothermic decomposition of the hydroxide but also, by the formation of a protective structure.

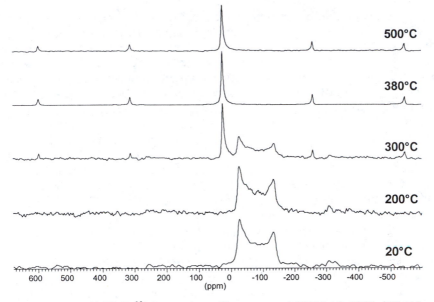

Figure 5. MAS NMR ^{25}Mg spectra of heat-treated EVA8-Mg(OH)$_2$/FB415 versus temperature

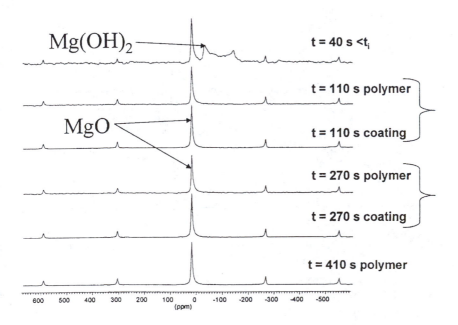

Figure 6. MAS NMR ^{25}Mg spectra of EVA8-Mg(OH)$_2$ versus time (cone calorimeter conditions)

No significant difference is observed by ^{13}C NMR between the formulations with and without FB415 (typical example on Figure 7). Bands around 32 ppm can be assigned to polyethylenic chains of the polymer and the region between 10 and 50 ppm correspond to aliphatic species (8). The broad bands centered about 130 ppm can be assigned to several types of aromatic and polyaromatic species (9)(10). Finally, the bands observed around 166 ppm can be assigned to magnesium carbonate (11).

After ignition, a coating of MgO ceramic is formed which can protect the underlying polymer (polyethylenic links observed until 410 s). The coating also contains a hydrocarbon structure, constituted of aliphatic and polyaromatic species formed by the degradation of the upper layer of EVA and/or by reaction of its degradation products. At the end of the combustion, the residue contains only magnesium carbonate and magnesium oxide. All "fuel" (polymer links protected by the protective ceramic) is consumed.

^{11}B NMR characterizes the evolution of the boron species in the condensed phase (Figure 8). Under the protective layer, no modification of FB415 is observed (the simulation of the spectra does not show significant modifications of boron). In the coating, FB415 is degraded and/or reacts. This result suggests an action of FB415 in the condensed phase. The simulation of the spectra support 4 different types of boron : 2 kinds of BO_3 units and 2 kinds of BO_4 units. Figure 9 shows the evolution of these units versus time. In a first step,

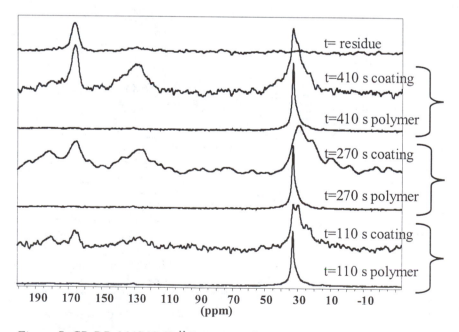

Figure 7. CP-DD-MAS NMR ^{13}C spectra of EVA8-Mg(OH)$_2$ versus time (cone calorimeter conditions)

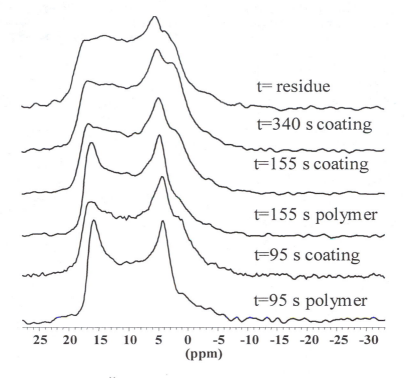

Figure 8. MAS NMR ^{11}B spectra of EVA8-Mg(OH)$_2$/FB415 versus time (cone calorimeter conditions)

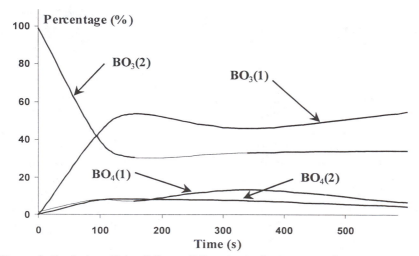

Figure 9. Evolution of the different BO$_x$ units in the protective layer versus time.

BO$_3$(2) (this unit is the major constituent of FB415) rapidly decreases during the heating of the formulation. Concurrently, another BO$_3$ is formed (BO$_3$(1)) which becomes the major specie during the combustion of the materials. The two BO$_4$ units are minor constituents. BO$_4$(1) comes from an impurity of FB415 and BO$_4$(2) from the degradation and/or the reaction of FB415 during the combustion of the material (12).

Figure 10 shows that B$_2$O$_3$ can be formed during the degradation of the formulation EVA8-Mg(OH)$_2$/FB415. The quadrupolar parameter of B$_2$O$_3$ is the same as BO$_3$(1). BO$_3$(1) can then be assigned to the formation of boron oxide during the flaming degradation of EVA8-Mg(OH)$_2$/FB415.

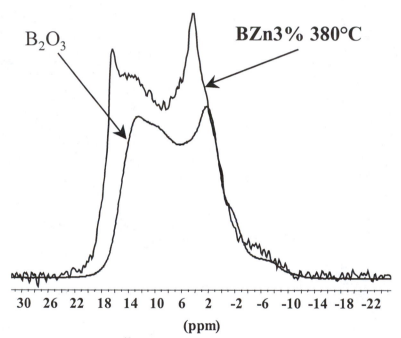

Figure 10. Simulation of ^{11}B NMR spectrum of EVA8-Mg(OH)$_2$/FB415 (3 wt.-% FB415 and noted BZn3% on the figure) heat-treated at 380°C using the quadrupolar parameter of B$_2$O$_3$

The simulation of the ^{11}B NMR spectra (Figure 11) shows that no B$_2$O$_3$ is formed during the thermal degradation of a formulation EVA8-FB415 but that two kinds of BO$_4$ units are detected (BO$_4$(1) and BO$_4$(2)). At 380°C, EVA8 is degraded evolving acetic acid, it can be assumed that the presence of BO$_4$(2) results from the reaction of the evolving acetic acid and FB415. In order to confirm this assumption, FB415 and acetic acid are mixed together at room temperature and the solid particles are filtered. ^{11}B NMR analysis (Figure 12) shows that BO$_4$(2) is formed with boron oxide. It may be then proposed, according to a previous work of Touval et al (13), that the BO$_4$(2) unit is due to the formation of a "vitreous ester" such as CH$_3$COO-B-(O-)$_3$.

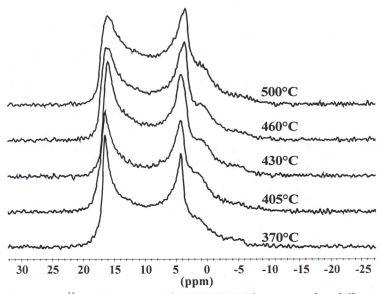

Figure 11. ^{11}B NMR spectra of EVA8-FB415 heat-treated at different temperatures

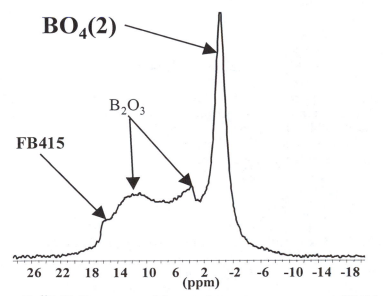

Figure 12. ^{11}B NMR spectrum of the reaction at room temperature FB415 and acetic acid.

The mode of action of Mg(OH)$_2$/FB415 in EVA8 can be described as follows :

- Endothermic decomposition of magnesium hydroxide during the heating of the material which leads to the increase of the time to ignition.
- Formation of a protective MgO ceramic. Its action is reinforced by FB415.
- FB415 reacts with the other constituents of the formulation which leads to the formation of boron oxide and vitreous esters. The boron oxide can play the role of binder which reinforces the efficiency of the MgO-based protective coating.

Conclusion

In this work, it is shown that zinc borate can act as a synergistic agent in combination with EVA-Mg(OH)$_2$. It is explained by the endothermic decomposition of Mg(OH)$_2$, which increases the time to ignition, and by the formation of a protective MgO-based ceramic. Zinc borate degrades into boron oxide which plays the role of a binder in the MgO-based ceramic, reinforcing the protective effect.

Acknowledgment

The authors are indebted to Mister Dubusse and Mister Noyon from CREPIM for their skilful experimental assistance in cone calorimeter experiments. NMR experiments were made in the common research center of the University of Lille, Mister Bertrand Revel is acknowledged for helpful discussion and experimental assistance.

References

(1) Hornsby, P.R.; Watson, C.L. *Plastic and Rubber Processing and Applications*, **1986**, *6*, 169-173.
(2) Shen K.; Ferm, D.F. *In : "Proceedings of 8th Recent Advances in Flame Retardancy of Polymeric Materials"*, Lewin M. Ed., BCC (Publisher), Stamford, **1997**.

(3) Bourbigot, S.; Le Bras, M., Leeuwendal, R.; Shen, K.K.; Schubert, D. *Polym. Deg. & Stab.*, **1999**, *64*, 419-425.

(4) Bourbigot, S.; Carpentier, F.; Le Bras, M.; Fernandez, C. *In : "Polymer Additives"*, Al-Malaika, S., Golovoy, A., Wilkie, C.A. Eds., Blackwell Science, London, **2000** in press.

(5) Amoureux, J.P.; Fernandez, C.; Carpentier, L.; Cochon, E. *Phys. Stat. Sol.*, **1992**, *132*, 461-471.

(6) Costello, C.A.; Schultz D.N. *In : "Kirk-Othmer Encyclopedia of Chemical Technology"*, 4th Edition, Wiley Interscience (Publisher), New-York, **1986**, pp. 349-381.

(7) Rothon R.N. *In : "Particulate-Filled Polymer Composites, Chapter 6 : Particulate Fillers used as Flame Retardant"*, Rothon, R.N. (Editor), Longman (Publisher), Harlow, **1995**.

(8) Duncan, M. *J. Phys Chem. Ref. Data*, **1987**, 16, 125-151.

(9) Maciel, G. E.; Bartuska, V. J.; Miknis, F. P. *Fuel*, **1979**, 58, 391.

(10) Bourbigot, S.; Le Bras, M.; Delobel, R. Trémillon, J.-M., *J. Chem. Soc., Faraday Trans.*, **1996**, 92(18), 3435-3444.

(11) Carpentier, F.; Bourbigot, S.; Le Bras, M. Delobel, R., *Polym. Deg. & Stab.*, **2000**, in press.

(12) Carpentier, F. *Thesis dissertation*, University of Lille, Lille, **2000**.

(13) Touval, I. *In : "Encyclopedia of Chemical Technology"*, 4th Ed., 10, pub. Wiley Interscience, New-York, **1985**, pp. 941-943.

Chapter 15

Optimization of Processing Parameters for Fire-Retardant Polymer-Based Formulations

L. Cartier, F. Carpentier, M. Le Bras, F. Poutch, and R. Delobel

ENCL, GéPIFREM, UPRES EA 2698, BP 108, F 59652 Villeneuve d'Ascq, France

This work deals with the effect of the processing parameters on the flame retardant and the rheological properties of two ethylene-vinyl acetate (EVA) copolymer-based formulations. The two formulations are based on both the concept of intumescence and the concept of ceramitization. For the intumescent formulation, the polymer degradation, particularly the thickening caused by the cross-linking of the polymer chains relating to processing conditions, is discussed. Finally, a principal component analysis allows the proposal of a relationship between processing and properties variables, and shows that heat transfer is responsible for the thickening of the polymeric material when high mechanical strains reduces the delay for this behavior. For the formulation that creates a ceramic-like residue, a first attempt is made to explain the difference in flame-retardant properties observed in the mixing and extrusion process.

This research includes the development of ethylene-vinyl acetate copolymer (EVA)-based formulations utilized in many industrial applications and particularly in low voltage cable and wire manufacture (*1-3*). Unfortunately, these polymers are frequently flammable and so it becomes important to reduce their flammability to increase their applications. The typical flame-retardant systems are based on halogenated compounds, which generate toxic by-products during their combustion. Thus, it is of importance to develop new halogen-free flame-retardant (FR) materials.

One way to solve the problems consists of using additives in order to build-up an intumescent structure, i.e. a cellular charred layer on the surface which protects the underlying polymeric substrate from the action of heat flux or flame. Our laboratory has recently developed an intumescent formulation: EVA8/APP/PA6 with 40 wt.-% additive loading. The EVA8 is the polymer matrix, ammonium polyphosphate the acid source and blowing agent, and polyamide 6 the carbonization agent (4). This formulation has not shown any interaction of the additives during processing.

A second approach of developing new FR materials consists of identifying systems that are able to form a vitreous protective layer over the polymer matrix and show endothermic reactions which absorb the flame energy and limit the degradation process. The layer acts as an insulator and a mass transport barrier hindering the escape of very often flammable volatile combustion products. For this purpose, the formulation: EVA19/Al(OH)$_3$/FBZB with 65% additives was evaluated. The EVA19 is the polymer matrix, aluminum hydroxide the mineral filler that decomposes into Al_2O_3 and water via an endothermic reaction, and zinc borate (FBZB) the synergistic agent and smoke suppressant (5).

The change in viscosity of the formulations influences their transport during processing. It is important to ensure that the laboratory conditions correlate with the actual industrial processing parameters. Processing parameters, such as, residence time, temperature, flow rate, extrusion or mixing speed, could modify the rheological properties of the formulations (6, 7). Therefore, they have to be optimized. The rheological modifications are due to the thermo-mechanical treatments of the material during their processing, which subsequently influence the FR properties of the formulations. Indeed, it is now well-known that the first step of the modification of the polymer matrix (i.e. formation of unsaturated groups and intermolecular crosslinks (8, and references therein) can have an effect on char generation by the matrix degradation (9). This additional char formation is frequently accompanied by a reduction in yield and/or rate of formation of volatile fuels. Moreover, the char can act, in addition to the protective material formed from the additives, as a physical barrier against heat transmission and oxygen diffusion (10).

A Brabender roller mixer and a counter-rotating twin-screw extruder were used to demonstrate the influences of residence time and temperature on both the FR performances and the rheological behavior of the formulations. The aim of the principal component analysis is to identify the respective effects of mechanical work and heat on the degradation process of the polymer during its processing and, to correlate their effect on the FR properties of the formulations.

Experimental

Materials

EVA8 (Evatane) and EVA19 (Escorene Ultra) with 8% and 19% (wt/wt) vinyl acetate content were supplied by Elf-Atochem and Exxon Chemicals respectively. Ammonium polyphosphate (APP), $((NH_4PO_3)_n$, n=700, Exolit 422) was supplied by Hoechst. Polyamide 6 (PA6) was obtained from Nyltech. Aluminum hydroxide $(Al(OH)_3)$ was supplied by Alcan (ATH SF-07/E) and zinc borate (Firebrake FBZB: $2ZnO.3B_2O_3.3.5H_2O$) was obtained from US Borax.

Formulations were processed using a Brabender Plasti Corder PL2000 roller mixer and a Brabender DSK 42/7 intermeshing counter-rotating twin-screw extruder. Limiting oxygen index (LOI) (*11*) was measured using a Stanton Redcroft instrument on sheets (120x10x3 mm³). The UL 94 test (*12*) was carried out on sheets (125x13x1.6 mm³). Rheological measurements were carried out on samples (25x25x1.6 mm³) using a Rheometric Scientific Ares-20A thermal scanning rheometer in a parallel plates configuration. The conditions of viscosity testing were: strain value (3°), angular rate (0.5 rad.s^{-1}), heating rate (20°C/min), force exerted on the samples (0.5 N).

Solid state NMR experiments were performed. ^{13}C NMR measurements were carried out on a Brucker ASX100 at 25.2 MHz (2.35 T) with magic angle spinning (MAS), high power ^1H decoupling (DD) and ^1H-^{13}C cross polarization (CP) using a 7 mm probe. A repetition time of 10 s and a spinning speed of 5 kHz were used. ^{27}Al and ^{11}B NMR measurements were performed on a Bruker ASX400 at respectively 24.5 MHz and 128.3 MHz (9.4 T) with MAS using a 4 mm probe. A repetition time of 10 s and a spinning speed of 15 kHz were used.

Principal components analysis

The principal components analysis was carried out using the STAT-ITCF computational method (Institut Technique des Céréales et des Fourrages, Paris). It is a descriptive method: from the obtained data, the relationships between the variables are derived directly from the computations; the neighboring variables may be explained by the existence of a relationship between two variables and that orthogonal positions may corresponds to the absence of relationships.

In this study, ten variables were used for the drawing of the correlation circles. Six centered variables which are: temperature gradient, GT; set temperature, Tc; thickening temperature, TE; heat required to mix the material, Q; mixer's rotor speed, V; and mechanical work required to mix the material, WMEC; and four supplementary "guest" variables that are not used in the computation of the principal axes: torque at the thickening point, CE; activation energy, E_A; constant, CO; torque at $1/T=0$, and from the calculation of the activation energy, time to reach the thickening point, TPS.

Results

Flame-retardant properties of the intumescent formulation

EVA8 exhibits poor heat resistance; its limiting oxygen index (LOI = 17 Vol.-%) is low and it is not classed (NC) by the UL 94 test. Thus, the polymer requires the incorporation of additives to achieve FR properties.

The intumescent formulation EVA8 (60 wt.-%)/APP (33 wt.-%)/PA6 (7 wt.-%) exhibits good flame-retardant properties after processing at 250°C using the twin-screw extruder at 50 rpm (1 rpm = 2πRad min^{-1}). It shows LOI = 30 Vol.-% (the experimental error on the LOI measurements is ±0.5 Vol.-%) and a V0 UL 94 rating, i.e. self-extinction and no dripping. The LOI values, considered as the FR properties vary as a function of both the screw speed and the processing temperature (Figure 1).

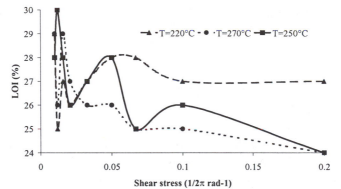

Figure 1: LOI values of EVA8/APP/PA6 processed using a counter-rotating twin-screw extruder at various temperatures and screw speeds.

In the low shear stress range (rotation ≤ 35 rpm), the LOI decreases when the processing temperature increases. The maximum value is reached when the rotation of the screw is 50 rpm in the 250-270°C processing temperature range.

Such behavior is not easy to understand at this stage of this study. It may be presumed that the variation of the measured LOI values results from chemical and/or physical changes of the intumescent formulation, arising from the thermo-mechanical effect of the process. It is obvious that such transformations modify the formulation's flow properties during the process. Therefore, it is necessary to study the formulation's rheological variations as a function of time and temperature to optimize the FR properties and to insure that the material can be processed in the optimum manner.

Flame-retardant properties of the ceramic-like formulations

The flame-retardant properties of EVA19 (LOI = 22 Vol.-% and NC by the UL 94 test) are significantly improved at 65 wt.-% total loading of $Al(OH)_3$ (LOI = 44 Vol.-% and V1 UL 94 rating). These properties are reinforced by substituting a small amount of aluminum hydroxide by zinc borate which reveals a synergistic effect for the EVA19/Al(OH)$_3$/FBZB formulations in the case of both a mixing and an extrusion process (Figure 2). This effect is optimum at 5 wt.-% of FBZB. Moreover, in this last case and for smaller ZB content, a V1 rating is still achieved whereas a higher FBZB content leads to a NC rating.

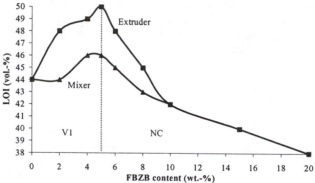

Figure 2: LOI values of EVA19/Al(OH)$_3$/FBZB (65 wt.-% loading) processed by using both the extruder and mixer versus the substitution in FBZB.

The LOI values strongly depend on the process. Indeed, the samples processed using the mixer at 190°C, 7 min residence time and 40 rpm screw speed lead to an optimal LOI of 46 Vol.-% which corresponds to a synergistic effect of 2 Vol.-%. The values obtained from extruded samples are generally higher and a maximum performance (LOI = 50 Vol.-%, synergistic effect of 6 Vol.-%) is reached at a processing temperature of 190°C, 7 min residence time and 40 rpm screw speed.

The explanation for this difference in LOI values under comparable processing conditions for the mixer and extruder could be due to fundamental differences between the mixing mechanisms: 1) mixing time; the mixing time in the mixer is about three times longer than in the extruder, 2) shear stress strength; for the same shear rate, the mechanical constraint is much higher in the extruder, 3) the presence of oxygen in the system; in the case of an extrusion process, the pressure gradient along the screw retains the air near the feeding zone while in the mixer the air is always in contact with the polymer formulation. This evolution of the two systems versus the zinc borate content may be correlated to the rheological and/or chemical modifications that are investigated in this paper.

Rheological variation of EVA8 versus the processing conditions

If chemical transformations of the material result in a significant change of the flow properties, they can be studied using both the mixer or the extruder at processing parameters such as torque, temperature, viscosity, and the activation energies. The two factors susceptible to the influence of the polymer's chemical degradation during its processing are the mechanical and the thermal constraints on the material (*13*). The extruder allows good shear stress control but it is quite impossible to control the temperature change of the molten polymeric material during the process, and to correctly measure this variable. Consequently, the material's viscosity may not be deduced from the material flow in dies. A roller mixer allows a nearly correct measurement of the two variables. The Brabender torque measured using a roller mixer (assumed to be a cylinder rheometer) may be related to the viscosity η of the thermoplastic material by: Torque $= K \eta = K \eta_0 \exp (E_A/RT)$
(with η_0 the pre-exponential factor and E_A the activation energy) when the temperature range is not too large (*14, 15*).

Plots of the logarithm of the torque versus T^{-1} give values of $K\eta_0$ and of E_A which are characteristic constants of a thermoplastic material; change of these values points out the modifications of the material. Figure 3 shows the change of EVA8 during the mixing process under air when the temperature of the melt increases (heating rate: $5°C$ min^{-1}). Linearity in the first zone corresponds to the thermoplastic behavior of EVA8. Change of the slope at about $200°C$ may be explained by the chemical change of EVA8 to form unsaturated groups with evolution of acetic acid, the decomposition product of vinyl acetate groups. In the range $205-220°C$ (zone 2), the degraded material is thermoplastic. In the $220-250°C$ region (zone 3), this linearity is not observed; a relative increase of the viscosity takes place with a computed difference in viscosities of 300 Pa.s.

This phenomenon can be explained by a cross-linking effect of the polymer, the increase in torque as a function of time being a characteristic of a thermoset resin. Finally in the last zone, the modified resin again shows the behavior of a thermoplastic. This shows that the chemical degradation modifies the flow properties of the material.

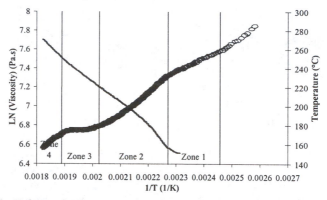

Figure 3: Values of viscosity as a function of the processing temperature (heating rate: 5°C/min) using a mixer (rotation speed: 50 rpm).

The variation in viscosity is observed when two consecutive temperature gradients are applied (respectively 5°C min[-1] and -5°C min[-1]) to the material (Figure 4). The two viscosity curves are superimposed when the temperature is above about 245°C. Thus, the irreversible chemical degradation of the EVA8 ends at 245°C after the application of the first temperature gradient.

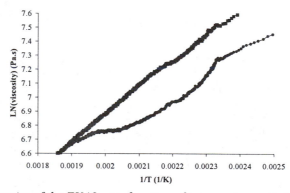

Figure 4: Viscosity of the EVA8 as a function of temperature, at mixer speed of 50 rpm. Two successive temperature gradients were applied between 150°C and 260°C and between 260°C and 150°C.

EVA8 mixing under nitrogen reveals that chemical modifications of the material are identical to those observed previously with the temperature gradient under air. This demonstrates that the modification of the EVA8 is not related to a chemical degradation due to the oxidation of the material; the chemical modification of the EVA8 is initiated by the thermo-mechanical constraints exerted on the material.

Chemical modification in the ceramic-like formulations

The processing parameters (temperature of 190°C, 7 min residence time and 40 rpm screw speed) used for the formulations EVA19/Al(OH)$_3$/FBZB processed using the mixer and the extruder are much less severe than those chosen for EVA8 (Figure 2). A degradation of EVA19 that causes the formulation to thicken is not expected. Indeed, ^{13}C NMR measurements confirm that EVA19 involved in the formulations is not modified during the mixing or the extrusion process.

The spectra of ^{27}Al NMR also show no differences between the Al(OH)$_3$ alone and the formulations made in the mixer and the extruder. The aluminum hydroxide is not modified during processing and does not interact with the other constituents of the formulation EVA19/Al(OH)$_3$/FBZB.

The ^{11}B NMR spectra of FBZB and the formulations processed using the mixer are very similar, implying that there is no major modification of FBZB during the mixing process. However, concerning the extrusion process, an evolution of the peaks is observed which corresponds to degradation of FBZB. The simulation of the spectra allows a proposal of two species with isotropic chemical shifts of about 1 and 18 ppm, which are assigned to BO$_4$ and BO$_3$ units respectively (16-18). The ratio BO$_3$/BO$_4$ versus FBZB content changes and is optimized at 5 wt.-% FBZB, which indicates that FBZB transforms during the extrusion process (Figure 5).

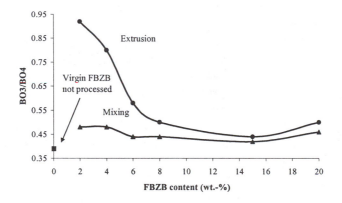

Figure 5: BO$_3$/BO$_4$ ratio versus FBZB content of the EVA19/Al(OH)$_3$/FBZB formulation after mixing and extrusion.

Rheological variation of EVA19/Al(OH)$_3$/FBZB versus temperature

The work on the intumescent formulations reveals that the chemical transformations of EVA8 during processing directly affect its rheological behaviour and influence the FR performance of the material. Since chemical modification of FBZB is demonstrated, the rheological fluctuations of the ceramic-like formulations need to be evaluated (Figure 6).

The viscosity measurements are performed in the temperature range from 200°C to 500°C for the formulations containing different amounts of FBZB and processed using the mixer and the extruder with similar conditions. Since no change is observed above 360°C, Figure 6 shows the formulations' rheological variations at low temperatures. The zinc borate content has a significant influence on the viscosity for all of the processes that have been utilized. Indeed, the viscosity increases to reach a maximum for a value of 5 wt.-% of FBZB before decreasing with the addition of larger zinc borate quantities in the formulations. However, it is important to note that the extrusion process leads to a higher viscosity compared to that obtained from the mixer. These observations should help understand the FR performance evolution which depends on both the substitution in FBZB of the formulations EVA19/Al(OH)₃/FBZB and the two different processes.

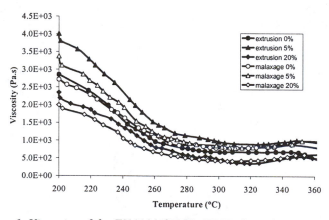

Figure 6: Viscosity of the EVA19/Al(OH)₃/FBZB formulations (mixing and extrusion) with various FBZB substitution levels, versus temperature.

Discussion

Intumescent formulation EVA8/APP/PA6

In this section, we will simply assume that the rheological and FR performances of the intumescent material are explained only in the EVA8 changes. The presence of the additives in the molten matrix does not significantly affect its properties during processing.

The chemical modification of the EVA8 is not due to its thermal oxidation, but rather to a thermo-mechanical-radical effect which relates to either the heating of the material or the mechanical work. In order to optimize the processing conditions of EVA8-based formulations, the influence of both the thickening mechanism parameters and exposure time on the constraints have to be identified. For this purpose, the principal components analysis is performed following a procedure developed by Heermann (*19*).

A part of the required mechanical work to mix the material is converted into heat that depended on shear stress rate. The difference between the set and processing temperatures becomes smaller when the latter is higher. The thermoplastic behavior of the material where viscosity decreases with temperature generates less friction at high temperature, thus explaining the heat change versus shear stress. In the analysis, the variable Q and WMEC are systematically corrected for this effect.

The relationships between all of the variables are then tested. Two significant correlation circles are computed (Figures 7-a and 7-b) using three principal axes. Axes 1 and 2 correspond respectively to the heat data (Tc, TE and Q) and the mechanical and thermal constraints that modify the viscosity of the formulation (GT and WMEC). Axis 4 corresponds to the global effect of less represented variables (GT, V and WMEC).

The activation energy E_A, characteristic of the material's chemical transformations (Figure 7), increases with the heat Q. The chemical modification of the material is favored by a large import of heat thus, a high setting temperature. The increase in the activation energy related to Q causes a decrease in viscosity according to the viscosity law. Logically, the torque CE (which is proportional to the decrease in viscosity) decreases when the activation energy E_A increases. Alternatively, the activation energy decreases when the rotor speed increases. Indeed, this rheological behavior of the material, where viscosity decreases with high rotor speed, is predominant over the decrease in viscosity related to temperature. It seems that at high rotor speeds, the thickening process is not favored or more precisely, it is hidden due to the breaking of the crosslinked chains as soon as they are generated.

The activation energy is independent of the thickening time TPS, the temperature gradient GT and, above all, the mechanical work WMEC. In addition, the thickening time is correlated to both the mechanical work and the temperature gradient. Therefore, the thickening process is related to the heat provided to the system and the application time of the thermal constraints, which is the residence time in the mixer.

The mechanical work increases with the thickening time and decreases when the rotor speed increases (Figure 7). As discussed previously, the high rotor speed partially hides the thickening process due to the polymer's rheological characteristics. The result is a decrease in viscosity and/or the breaking of the polymeric chains.

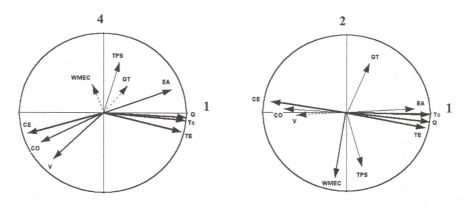

Figure 7: Correlation circles for axes 1, 2 and 4 (the degree in consideration of the variables is presented in decreasing order as followed: heavy, thin and dashed lines respectively).

This study demonstrates that the mechanical constraints play two roles in the system. The first may be explained by the generation of free radicals that initiate the crosslinking reaction when the application time of the mechanical constraints or residence time is long enough. The second is the large friction to the material that leads to an increase in the processing temperature.

To resume, the thermal phenomenon is responsible for the thickening of EVA8. The shear stress accelerates the initiation of the phenomenon at temperature below 200°C. Therefore, it is possible to rheologically modify a material according two different procedures: a) The material is exposed to a thermal constraint for a period of time which corresponds to a major part of the residence time in the extruder; and b) The material is exposed to a high mechanical constraint that generates a local elevation of temperature, which is the initiating factor of the process.

The evolution of the FR performance versus the processing conditions (temperature and shear stress using the extruder) may now be explained from these results. The residence time of the polymeric materials in the extruder is low (from 30 to 120 s., a function of the screw rotation speed). At 220°C and low screw rotation speeds, the degradation of the vinyl acetate groups of EVA8 is not complete when the high temperature allows degradation. In a fire, release of the acetate allows the evolution of acetic acid, which has a dilution effect in the flame. Temperatures higher than 220°C in this rotation range give an easily flammable unsaturated polymer. Using a screw rotation of 50 rpm at all temperatures allows intermolecular crosslinking of the polymer chains to occur whose charring gives comparatively high FR performances. Finally, in the highest rotation speed range, the breaking of the polymeric chains occurs which leads to easily flammable low molecular weight species.

Ceramic-like formulation EVA19/Al(OH)$_3$/FBZB

In the previous sections, it is shown that the rheological and FR properties of the EVA19/Al(OH)$_3$/FBZB formulations depend on both the zinc borate transformation and the process. Indeed, the later influences the FR performances since different maximum LOI values of 46 Vol.-% and 50 Vol.-% are found for mixing and extrusion, respectively.

The mechanism of action of zinc borates in such formulations has been recently proposed (5). During the heating of the material, the polymer degrades and forms a cross-linked network and a carbonaceous char, aluminum hydroxide decomposes into Al$_2$O$_3$ and FBZB undergoes a transformation that leads to a vitreous protective coating made of a boron-based glassy phase which allows a better retention of the char. The mechanically stable vitreous coating formation is evidenced by the increase of the BO$_3$/BO$_4$ ratio that reaches a plateau with time. In the formulations discussed in this paper, the degradation of FBZB during the process is more important in the case of extrusion than mixing. The good LOI results of the extruded samples may be caused by this small degradation of FBZB. Boron oxide may act as an initiator and its incorporation (from 0.5 to 1 wt.-%) in the formulation during the process may increase the kinetics of boron oxide formation that would lead to better FR performances. The boron-based glassy coating may be also of a better quality. ^{11}B NMR studies at different degradation time of the formulations processed using the mixer and the extruder should be performed to confirm this hypothesis.

Also, the mechanical constraint, which is much stronger in the extruder, may reduce the particle size of the components of the blend. This behavior generally has an impact on the FR performance of the materials. Experiments with particle sizes of 0.5 μm for aluminium hydroxyde and 3 μm for FBZB (instead of respectively 0.7 μm and 7 μm used in this paper) are under evaluation. Particular attention will be paid on the impact of the smaller zinc borate particle sizes on the BO$_3$/BO$_4$ ratio.

The low temperature viscosity measurement is correlated to the LOI test Indeed, for both process, i.e. extrusion and mixing, the LOI values vary as a function of the viscosity measurements (Figures 2 and 6). In the LOI test, the samples are placed in a vertical position and the protective layer when formed and generated during the combustion may flow down due to the lack of resistance of the melted polymer phase supporting the protective layer. In other words, the support of the protective layer on the surface of the sample depends on the viscosity of the material. The higher is the viscosity the better is the LOI (20), thus explaining the high LOI values observed at a 5 wt.-% FBZB loading, which is the content that leads to the highest viscosities.

The two different processes that are used lead to the same UL 94 classification. The formulations with a total loading of FBZB below and above 5 wt.-% exhibit a V1 and a NC rank, respectively. In this case, the difference in the viscosity measurement between the mixing and the extrusion process is not large enough to compensate for the force exerted by the weight of the samples.

Conclusion

The study shows that FR Properties of intumescent EVA8/APP/PA6 formulations depend on their processing conditions. A relationship between high FR ratings and cross-linking (thickening) of the EVA8 phase in the material is discussed.

It is shown that the increase in viscosity can be assigned to the thermal treatment. Moreover, shear stress, i.e. mechanical constraint, plays two distinct parts; it allows the formation of the species which initiate the thickening process and contributes also to the breaking of intermolecular links and to the degradation of the polymer chains.

The difference in flame retardant properties for the ceramic-like formulations EVA19/Al(OH)$_3$/FBZB processed using the mixer and the extruder was discussed. In a first approach, it is shown that this difference may be explained by a decrease in the FBZB particle size due to higher mechanical constraints that occur in the extrusion process compared to the mixing process.

References

1. Ray, I.; Khastgir, D. *J. Appl. Polym. Sci.* **1994**, *53*, 297-307.
2. Ray, I.; Khastgir, D. *Plastics, Rubber and Composites Processing and Applications* **1994**, *22*, 37-45.
3. Fedor, A.R. *Int. Wire and Cable Symp. Processings* **1998**, 183-187.
4. Siat, C. Ph.D thesis, Université d'Artois, Lens, France, 2000.
5. Bourbigot, S.; Le Bras, M.; Leeuwendal, R.; Shen, K.K.; Schubert, D. *Polym. Deg. Stab.* **1999**, *64*, 419-425.
6. Seibel, S.R.; Papazoglou, E. *Annu. Tech. Conf.- Soc.Plast. Eng.* **1995**, *53*, 366-370.
7. Beecher, E.; Grillo, J.; Papazoglou, E.; Seibel, S.R. *Annu. Tech. Conf.- Soc.Plast. Eng.* **1994**, *52*, 1626-1630.
8. Pécoul, N. Ph.D thesis, USTL, Lille, France, 1996.
9. Balabanovich, A.I.; Schnabel, W.; Levchik, G.F.; Levchik, S.V.; Wilkie, C.A. In *"Fire Retardancy of Polymers - The Use of Intumescence"*, Le Bras M.; Camino, G.; Bourbigot, S.; Delobel, R. Eds., The Royal Society of Chemistry, Cambridge, UK, **1998**, 236-251.
10. Carty, P.; White, S. *Polymer* **1994**, *35*, 343-347.

11. Standard Test Method for Measuring the Minimum Oxygen Concentration to Support Candle-like Combustion of Plastics, ASTM D 2863-ISO 4589. * It is generally assumed that the higher the LOI value is the better is the FR behavior.
12. Tests for Flammability of Plastic Materials for Part in Devices and Appliances, *Underwriters Laboratory Safety Standard 94*, ASTM D 568-ISO 1210:1992.
13. Zaikov, G.E.; Goldberg, V. In *Encyclopedia of Fluid Mechanics*, Gulf Publishing Company, Houston, Texas, 1990, *9, 13*, 403-423.
14. Rothenpieler, A. In Mesure des Propriétés de Transformation des Hauts Polymères avec le PLASTI-CORDER Brabender, Réunion Commune *AFICEP-SPE-GFP*, Paris, 23 mars 1973.
15. Poutch, F. Mémoire CNAM (available at GEPIFREM, E.N.S.C.L), Lille, France, 1998.
16. Turner, G.L.; Smith, K.A.; Kirkpatrick, R.J.; Oldfield, R.J. *J. Magn. Reson.* **1986**, *67*, 544-550.
17. Bray, P.J.; Edwards, J.O.; O'Keefe, J.G.; Ross, V.F., Tatsuzaki, I. *J. Chem. Phys.* **1961**, *35*, 435-442.
18. Kriz, H.M.; Bishop, S.B.; Bray, P.J. *J. Chem. Phys.* **1968**, *49*, 557-561.
19. Heermann, E.F. *Psych.* **1963**, *28*, 161-172.
20. Carpentier F. Ph. D. thesis, USTL, Lille, France, 2000.

Chapter 16

Fire Retardancy and Thermal Stability of Materials Coated by Organosilicon Thin Films Using a Cold Remote Plasma Process

C. Jama[1,2], A. Quédé[1,2], P. Goudmand[1], O. Dessaux[1], M. Le Bras[2],
R. Delobel[2], S. Bourbigot[3], J. W. Gilman[4], and T. Kashiwagi[4]

[1]Laboratoire de Génie des Procédés d'Interactions Fluides Réactifs-Matériaux,
U.S.T.L., BP 108, 59652 Villeneuve d'Ascq Cedex, France
(Telephone: +33 (0)3 20 33 63 11; email: jama@univ-lille1.fr)
[2]Laboratoire de Génie des Procédés d'Interactions Fluides Réactifs-Matériaux, Ecole
Nationale Supérieure de Chimie de Lille, BP 108, 59652 Villeneuve d'Ascq Cedex,
France (email: michel.le-bras@ensc-lille.fr)
[3]Laboratoire de Génie et Matériaux Textiles, ENSAIT, BP 30329,
59056 Roubaix Cedex 1, France
[4]Materials Fire Research Group, Fire Science Division, National Institute
of Technology, Gaithersburg, MD 20899

This investigation is concerned with the thermal stability of organosilicon thin films obtained from the polymerization of 1,1,3,3-tetramethyldisiloxane ($H(CH_3)_2Si$-O-$Si(CH_3)_2H$) (TMDS) monomer doped with oxygen using the cold remote nitrogen plasma (CRNP) process. The role played by oxygen added to TMDS monomer during the polymerization process on the thermal stability of the deposited films is clearly evident. The thermal degradation behavior of deposits under pyrolytic and thermo-oxidative conditions shows that the residual weight evolution with temperature depends on the chosen atmosphere. Higher amounts of a solid residue are obtained in air, which demonstrates that atmospheric oxygen participates in the degradation mechanism leading to the formation of a more thermally stable surface product. In terms of fire retardancy performance, it is shown that the deposited films are efficient fire retardant coatings. The CRNP process was used to deposit thin films on a polyamide-6 (PA-6) and polyamide-6 clay nanocomposite (PA-6 nano) substrates. Limiting Oxygen Index (LOI) values obtained for PA-6 show only a slight increase compared to PA-6 nano. However, the LOI obtained for the coated PA-6 nano are drastically improved. The peak rate of heat release (Pk. RHR) of the coated materials is also decreased by about 30% in comparison to the virgin polymer.

Polymers are used in many fields, such as housing materials, transport and electrical engineering applications. Due to their chemical constitution, these polymers are easily flammable and so, flame retardancy becomes an important requirement for many of them. This can be obtained in several ways such as incorporation of additives (the classical method), chemical modification of the macromolecule, or by surface modification (γ-rays radiation) *(1-3)*. The first process provides good flame retardancy properties and many flame-retardants are commercially available. However, additive loading can reduce the thermal and mechanical properties of the polymer or lead to ecological problems (e.g. incineration of polymers loaded with halogenated flame-retardants) *(4)*. Chemical modification of the polymer is generally not easy to do on the industrial scale. The third possibility seems to be attractive but much work remains to be done.

We investigate another route for surface modification by laying down a thin film on the polymer using the plasma assisted polymerization technique. Plasma enhanced chemical vapor deposition (PECVD) is one of the widely used processes for the deposition of thin films. Nevertheless with PECVD, there is always some substrate damage induced to sensitive substrates by high-energy particles bombardment from the plasma. Among the numerous varieties of plasma assisted polymerization techniques, Remote PECVD (RPECVD) has recently attracted considerable interest to overcome this disadvantage. Our laboratory has developed a plasma deposition process induced by cold remote nitrogen plasma (CRNP). In this process the substrate is removed from the plasma as the plasma and the reaction chamber are spatially separated *(5-9)*. The main reactive species of the CRNP are nitrogen atoms in the ground electronic state and electronically and vibrationally excited nitrogen molecules. The CRNP is then a non-ionized reactive zone and is characterized by an important thermodynamic non-equilibrium state.

CRNP assisted polymerization of organosilicon compounds is an interesting preparation technique of thin polysiloxane-based films. This polymer is known to have good thermal stability. The CRNP polymerization mechanism and the structure of the organosilicon polymer are described by Supiot *et al.*(5, 6). Specific industrial aspects of the CRNP assisted polymerization have been protected by industrial patents *(10-12)*. Previously, we have presented preliminary results on fire retardancy of PA-6 by such process by reaction with 1,1,3,3-tetramethyldisiloxane (TMDS) monomer $(H(CH_3)_2Si-O-Si(CH_3)_2H)$ doped with oxygen *(13)*. In this work, the CRNP process is used to deposit thin films on polyamide-6 (PA-6) and polyamide-6 clay nanocomposite (PA-6 nano, clay mass fraction of 2 wt. %) substrates. PA-6 nano is synthesized according to the procedure described by Gilman et al *(14)*. The thermal behavior of both PA-6 and PA-6 nano have been discussed by Dabrowski *et al.* *(15)*. Here, we present results on the role played by oxygen added to TMDS monomer during the polymerization process on the thermal stability of the deposited films. The aim is to study the thermal degradation behaviors of deposits under pyrolytic and thermo-oxidative conditions.

The flame retardancy of the coated materials is evaluated using limiting oxygen index (LOI) *(16)* and cone calorimetery *(17)* to obtain the rate of heat release (RHR), smoke emission and gases evolved during combustion versus oxygen addition to TMDS monomer.

Experimental

Materials: Raw materials are polyamide-6 (PA-6, as pellets, supplied by UBE), polyamide-6 clay nanocomposite (PA-6 nano, as pellets, supplied by UBE, clay mass fraction of 2 wt.-%). Sheets are then obtained using a Darragon press at 23°C with a pressure of 10^6 Pa. The monomer is 1,1,3,3-tetramethyldisiloxane 97% (TMDS), supplied by Aldrich Chemical Co.

Plasma device: The experimental setup of the CRNP assisted deposition reactor is described in Figure 1. Nitrogen is introduced under a pressure of 4.2 hPa at a flow rate of 1.8 slpm (standard liter per minute) in a quartz tube of 33 mm outer diameter. The discharge excitation is created by a coaxial coupling device at 2450 MHz. The gas containing excited species is extracted from the discharge zone by a primary pump. The reactive gas, TMDS mixed with oxygen, is injected into the reaction chamber through a coaxial injector at a distance of 1.5 m downstream from the discharge.

Samples are placed horizontally in the reactor and are coated successively on each face; an open-air operation separates these two steps. In order to obtain homogeneous deposits, the substrates are located at a distance of 10 cm from the injector inlet. The incident microwave power is maintained constant at 600 W. Oxygen flow rate in the range of 10 – 200 sccm (standard cubic centimeter per minute) is mixed to a constant TMDS flow rate of 5 sccm. A deposition time of 20 min is used for each surface.

Thermal analysis: Thermogravimetric analyses (TGA) are carried out from 20°C to 800°C at five heating rates (2.5, 5, 7.5, 10 and 12.5°C / min) under air flow (Air Liquide grade, $5\times10^{-7}m^3$/s measured in standard conditions) and at four heating rates (5, 7.5, 10 and 12.5°C / min) under nitrogen flow (N45 Air Liquide grade, $5\times10^{-7}m^3$/s measured in standard conditions) using a Setaram apparatus. Samples are placed in open vitreous silica pans. Thermal analysis is carried out on coatings in the powder form (~10 mg). The coating, as powdered material, is obtained by scraping the films deposited on aluminum plates. Precision on temperature measurements is ±1.5°C.

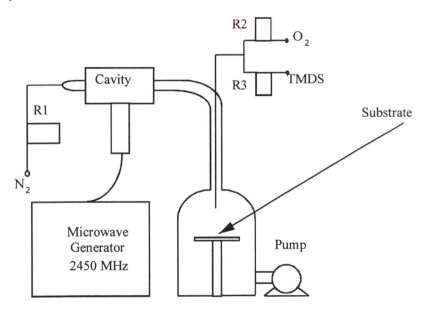

R : Regulators

Figure 1 : Experimental set-up of CRNP assisted deposition process.

Limiting Oxygen Index (LOI): LOI is measured using a Stanton Redcroft instrument according to the standard oxygen index test *(16)*.

Cone calorimetry: Samples $(2 \times 2 \times 0.3)\,cm^3$ are exposed to a Stanton Redcroft Cone Calorimeter under an external heat flux of 35 kW/m^2 which represents the heat flux found in the vicinity of solid-fuel ignition source. Conventional data (Rate of Heat Release (RHR), Volume of Smoke Production (VSP), ...) are investigated using software developed in this laboratory.

Results and Discussion

Effect of oxygen addition : Figure 2 gathers results on thermal stability under air at a heating rate of 7.5°C/min for deposits obtained on aluminum plates from TMDS mixed with oxygen at different flow rates ($\Phi(O_2)$). It shows that the coating is thermally very stable.

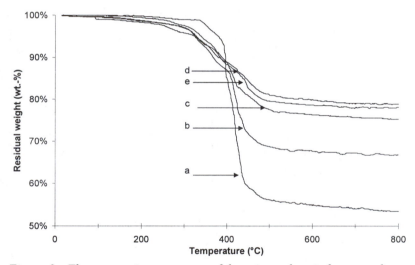

Figure 2 : Thermogravimetry curves of deposits under air for several oxygen flow rates: (a: 10 sccm, b: 30 sccm, c: 50 sccm, d: 150 sccm and e: 200 sccm).

At $\Phi(O_2)$ = 10 sccm, the deposit decomposes between 320 and 480°C and gives 53 wt.-% of solid residue at 800°C. By increasing $\Phi(O_2)$ from 10 to 200 sccm, it is of interest to note that the degradation process begins at lower temperature (around 280°C), whereas it finishes at similar temperature whatever the oxygen flow rate. Moreover, the residual weights sharply increase when $\Phi(O_2)$ increases from 10 to 50 sccm and vary very slightly at higher oxygen flow rate. These data clearly indicate that increasing oxygen flow during the polymerization promotes the formation of films more thermally stable in air. The structure of the organosilicon polymer evolution versus oxygen addition is well known (6). FTIR results show the presence in the film of Si-O-Si, Si(CH$_3$) in a polysiloxane-like structure and Si(OH) groups. For films deposited at $\Phi(O_2)$ = 50 sccm, the evolution of the film nature as function of the temperature shows a decrease in the intensity of Si(CH$_3$), which completely disappear at 800°C. Moreover, the structure of the film shifts from a polysiloxane like to a SiO$_2$-like structure by heating the deposit from 25°C to 800°C. These results corroborate the high thermal stability of the deposited coating.

Effect of heating rate

Thermo-oxidative condition: Thermogravimetric curves under air of the organosilicon deposit prepared at an oxygen flow rate of 10 sccm at five heating rates are presented in Figure 3. It appears that whatever the heating rate, the mechanism of degradation occurs in a multi-step process. On the other hand, it is of interest to note that the heating rate drastically influences the thermal behavior of the deposits.

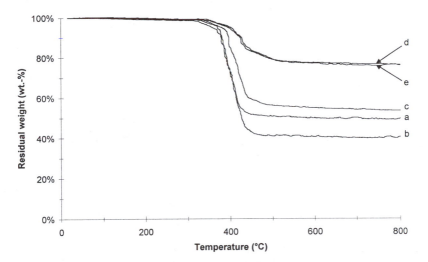

Figure 3 : TGA curves under air of organosilicon deposit prepared at $\Phi(O_2)$=10 sccm for several heating rates: (a: 2.5, b: 5, c: 7.5, d: 10 and e: 12.5 °C / min).

By increasing the heating rates from 2.5°C/min to 5°C/min we can observe that there is a decrease in the residual mass. For higher heating rates (from 5°C/min to 12.5°C/min) an increase of the residual mass is observed. This suggests that there is a participation of oxygen in the degradation process, which depends on the heating rate. One possible explanation is the existence of a conjugated multi-stage process. The simplest one is the following mechanism:

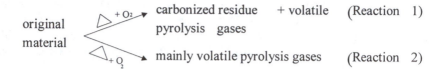

These two reactions are competitive. If the activation energy of reaction 2 is lower than that for reaction 1, then, by increasing the heating rate, the yield of a solid product will increase. If the mechanism seems simple, its mathematical modeling is not easy to solve. The computation of kinetics under thermo-oxidative condition of the films will be the subject of a future study to optimize the protective process.

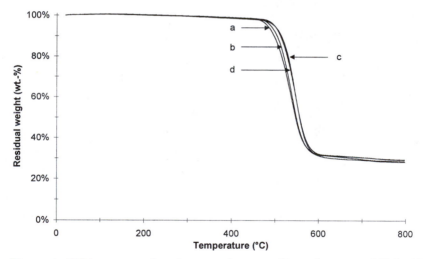

Figure 4 : TGA curves under nitrogen of organosilicon deposit at $\Phi(O_2)=10$ sccm for several heating rates: (a: 5, b: 7.5, c: 10 and d: 12.5 °C/ min).

Pyrolytic condition: Thermogravimetric curves under nitrogen of the organosilicon films at oxygen flow rate of 10 sccm at different heating rates are presented in Figure 4. Whatever the heating rate, only one step of degradation is observed. The deposit begins to degrade at 470°C; at 600°C, the residual mass is 30% wt.-% and it remains stable until 800°C. At lower heating rates the degradation begins earlier.

Comparison: Figure 5 shows thermogravimetry curves under nitrogen and under air for the deposit obtained at oxygen flow rate of 10 sccm. Residual weight evolution with temperature shows different behavior, depending on the atmosphere. Under air the deposit begins to lose mass at about 320°C, and at 800°C, 53 wt.-% of solid residue remains. Under nitrogen, the major step of degradation shifts to 470°C, and only 30% wt.-% of a solid residue remains at 800°C. The larger amount of the solid residue obtained in air demonstrates that atmospheric oxygen participates in the degradation mechanism, leading to a more thermally stable product. Further NMR analyses will give us a clear diagnosis on the nature of the residue obtained under nitrogen or under air.

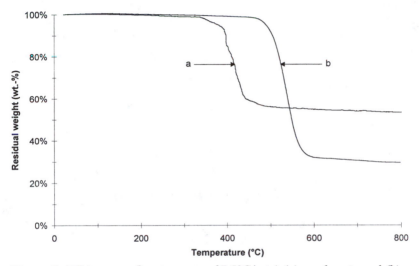

Figure 5: TGA curves (heating rate of 7.5°C/min) (a): under air and (b): under nitrogen of organosilicon deposit obtained at $\Phi(O_2)=10$ sccm.

Fire Retardancy performance

Figure 6 presents the LOI (Limiting Oxygen Index) values versus oxygen flow rate ($\phi(O_2)$) doping TMDS for virgin and coated substrates. The LOI for the PA-6 nano increase very dramatically when oxygen is present while there is only a small increase for PA-6. At $\Phi(O_2)=10$ sccm, the LOI for the PA-6 nano is 43± 2 vol % while that for the virgin PA-6 nano is 23± 1 vol %. Higher LOI values are obtained by increasing oxygen flow rate. The increase of LOI values with oxygen is in agreement with the fact that oxygen addition to the monomer

208

promotes the formation of stable films as clearly evidenced from thermal analysis.

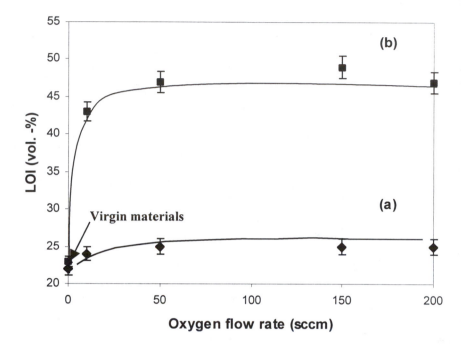

Figure 6 : LOI versus oxygen flow rate in the CRNP reactor for coated (a) : PA-6 and (b) : PA-6 nano

Figure 7 shows the LOI values evolution obtained for coated PA-6 and PA-6 nano versus the residual weight of organosilicon deposit obtained at 800°C under thermo-oxidative conditions. We can observe in the case of PA-6, that the thermal stability of the deposit influences the LOI values only slightly. However, in the case of PA-6 nano, the LOI values increase as the thermal stability of the coatings increase. This result gives evidence of a synergetic effect between the deposited film and the polyamide-6 clay nanocomposite. A difference in terms of adhesion between the films deposited on PA-6 and PA-6 nano could be an explanation of this effect. This means that more adherent films could be obtained on PA-6 nano. It could also be due to possible reactions between the film and the clay nanocomposite that could form more thermally stable films. FTIR analysis of the char from the LOI experiments indicate that a polysiloxane-like structure is obtained from the coated PA-6. In the case of coated PA-6 nano a SiO_2-like structure is formed. These results give evidence of synergism with the clay nanocomposite. Such an effect with PA-6 nano can be summarized as being due to a reaction between the deposited film and the clay nanocomposite, giving a thermally more stable system, and/or to an enhancement of the adhesion quality between the coatings and the PA-6 nano.

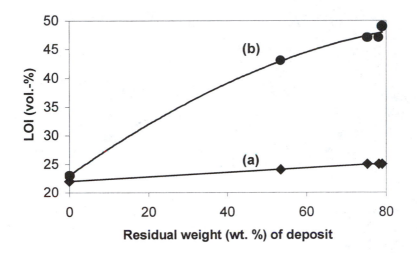

Figure 7 : LOI evolution of coated (a) : PA-6 and (b) : PA-6 nano at 10, 50, 150 and 200 sccm $\phi(O_2)$ rates versus residual weight of organosilicon deposit

Rate of Heat Release (RHR) curves of coated ($\phi(O_2)$ =50 sccm) and virgin PA-6 are presented in Figure 8. The shapes of the RHR curves are similar. The ignition time of the coated PA-6 is lower than that of virgin PA-6. However the peak RHR of the coated PA-6 (1300 kW/m²) is decreased by 30% in comparison to virgin PA-6 (1850 kW/m²).

The RHR curves of the coated PA-6 nano at $\phi(O_2)$ = 50 sccm and of the virgin PA-6 nano are presented in Figure 9. The ignition time of the coated PA-6 nano is close to that of virgin PA-6 nano. However the peak RHR of the coated PA-6 nano (820 kW/m²) is decreased by 25% in comparison to virgin PA-6 nano (1100 kW/m²).

The evolution of the maximum of RHR, THE (Total Heat Evolved), VSP, CO and CO_2 versus $\phi(O_2)$ are gathered in Tables I and II for PA-6 and PA-6 nano respectively. They show interesting results. The CO and CO_2 values of coated PA-6 nano are very similar to those obtain for the virgin material. For PA-6, CO values of the coated material are very high in comparison with the virgin sample. In the case of the CO_2, the contrary is observed. This means that the combustion process is different for the two polymers and seems to be more incomplete in the case of coated PA-6 polymer. It could also be due to a difference in term of selectivity between the reactions of complete combustion and those of CO production.

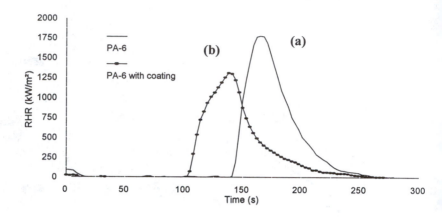

Figure 8: RHR values versus time for virgin (a) and coated PA-6 (b) at $\phi(O_2) = 50$ sccm.

After ignition, the coating forms a physical barrier and the RHR values are reduced. The lower value of ignition time obtained for coated PA-6 could be due to a delamination of the coating under heating inducing some defects, which may promote faster degradation during heating and then decrease the ignition time. The presence of "free radicals" species on the polymer surface resulting from the CRNP process can act as an initiator of the degradation process and, as a result, decrease the time needed to obtain inflammability conditions.

For PA-6 nano, the quite stable value of ignition time and the higher LOI values are consistent with a good adhesion quality of the film and with synergism with the clay nanocomposite FR mechanism, as suggested from the LOI behavior.

Table I. Evolution of the conventional cone calorimeter data versus oxygen for PA-6

	PA-6	*10 sccm*	*30 sccm*	*50 sccm*
RHR (kW/m²)	1850	1105	1197	1300
THE (kJ)	30	39	38	23
VSP (m³/s)	4.4 10⁻⁴	3.2 10⁻⁴	4.6 10⁻⁴	4.4 10⁻⁴
CO (ppm)	8.5	26	18	25
CO₂ (vol.-%)	0.130	0.094	0.099	0.098
Ignition time(s)	140	80	82	100

Table II. Evolution of the conventional cone calorimeter data versus oxygen for PA-6 nano

	PA-6 nano	10 sccm	30 sccm	50 sccm
RHR (kW/m^2)	1100	1051	1168	819
THE (kJ)	35	34	34	31
VSP (m^3/s)	5 10^{-4}	4.7 10^{-4}	4.2 10^{-4}	4.8 10^{-4}
CO (ppm)	6.9	2.6	9.3	11.3
CO$_2$ (vol.-%)	0.084	0.076	0.096	0.068
Ignition time (s)	81	83	80	93

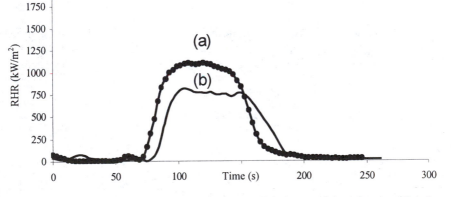

Figure 9: RHR values versus time for virgin PA-6 nano (a) and coated PA-6 nano (b) at $\phi(O_2) = 50$ sccm.

Conclusion

In this work, we have used the cold remote nitrogen plasma (CRNP) process to obtain thin protective films on a polyamide-6 (PA-6) and polyamide-6 clay nanocomposite (PA-6 nano) substrates. These films are obtained from polymerization of 1,1,3,3-tetramethyldisiloxane (TMDS) monomer doped with oxygen using CRNP process. Oxygen addition to TMDS monomer promotes the formation of thermally stable films. The thermal degradation behavior of the deposits under pyrolytic and thermo-oxidative conditions demonstrates that atmospheric oxygen participates in the degradation mechanism, leading to the formation of a more thermally stable products.

It is shown that the deposits are efficient fire retardant coatings. Limiting Oxygen Index values obtained for polyamide-6 nanocomposites are drastically improved. This result is probably due to a good adhesion quality of the film and/or to possible synergetic reactions between the deposited film and the clay nanocomposite. The rate of heat release (RHR) of the coated materials at 50 sccm flow rate of O_2 is also decreased by about 30% in comparison to the virgin polymer

Acknowledgements

The authors thank Miss S. Duquesne from ENSCL, Dr. Vadim Mamleev (Univ. Alma Aty, Khazakstan) for their helpful discussions and Mr. José Cardoso (Instituto Superior de Engenharia do Porto, Portugal) for technical assistance.

References

1. Lewin, M. *in Fire Retardancy of Polymers - The Use of Intumescence*, eds. Le Bras, M.; Camino, G.; Bourbigot, S.; Delobel, R. The Royal Society of Chemistry, Cambridge, U. K. , **1998**, p. 3-32.
2. Troitzsch, J. In *International Plastics Flammability Handbook, 2nd Edition*, ed. Troitzsch, J. Hanser Publishers, Munich, G, **1990**, p. 43-62.
3. Brossas, J. Techniques de l'Ingénieur, ed. Debaene, J.; Isra BL, Schiltigheim, Paris, F, **1998**, AM 3 237 p. 1-12.
4. Nelson, G. L. *Fire and Polymers II*, ACS Symposium Series 599, ed. Nelson, G. L. American Chemical Society, Washington, DC, **1995**, p. 1-28.
5. Supiot, P.; Callebert, F.; Dessaux, O.; Goudmand, P. *Plasma Chem. Plasma Proc.*, **1993**, *13*, 539-554.

6. Callebert, F.; Supiot, P.; Asfardjani, K.; Dessaux, O.; Goudmand, P.; Dhamelincourt, P.; Laureyns, J. *J. Appl. Polym. Sci.*, **1994**, *52*, 1595-1606.

7. Jama, C., Ph.D. thesis, USTL, Lille, F, **1995**.

8. Jama, C.; Asfardjani, K.; Dessaux, O.; Goudmand, P.; *J. Appl. Polym. Sci.*, **1997**, *64*, 699-705.

9. Dessaux, O.; Goudmand, P.; Jama, C.; *Surf. Coatings Technology*, **1998**, *100-101*, 34-44.

10. Callebert, F.; Dupret, C.; Dessaux, O.; Goudmand, P.; Intern. Pat. Application, WO 92/03591 (**1992**).

11. Callebert, F.; Supiot, P.; Dessaux, O.; Goudmand, P.; PCT/FR94/00149 (**1994**).

12. Caburet, L.; Asfardjani, K.; Dessaux, O.; Goudmand, P.; Jama, C.; FR 95/05333, **1995**.

13. Bourbigot, S.; Jama, C.; Le Bras, M.; Delobel, R.; Dessaux, O.; Goudmand, P. *Polym. Deg. Stab.*, **1999**, *66*, 153-155.

14. Gilman, J. W.; Kashiwagi, T.; Giannelis, E. P.; Manias, E.; Lomakin, S.; Lichtenhan J. D., Jones P. in *Fire Retardancy of Polymers -The Use of Intumescence*, eds. Le Bras, M.; Camino, G.; Bourbigot, S.; Delobel, R. vol. 203. Cambridge (UK): Royal Chem. Soc., **1998**, pp. 203-221.

15. Dabrowski, F.; Bourbigot, S.; Delobel, R.; Le Bras, M.; *Eur. Polym. J.,* **2000**, *36*, 273-284.

16. Standard test method for measuring the minimum oxygen concentration to support candle like combustion of plastics, ASTM D 2863/77: American Society for testing and materials: Philadelphia, PA, **1977**.

17. Babraukas, V. *Fire Mater.*; **1989**, *8*, 81-95.

Chapter 17

Phosphorus-Containing Fire Retardants in Aliphatic Nylons

Sergei V. Levchik[1], Galina F. Levchik[2], and Elena A. Murashko[2]

[1]Akzo Nobel Functional Chemicals LLC, 1 Livingston Avenue,
Dobbs Ferry, NY 10522
[2]Research Institute for Physical Chemical Problems, Belarussian University,
Leningradskaya 14, Minsk 220050, Belarus

An overview of technical literature from the past 10 years on use of phosphorus-containing compounds in aliphatic nylons showed significant effort in the search for new products providing a good balance of fire retardancy and physical properties. The authors' study of the fire retardant action of resorcinol bis(diphenyl phosphate) and bisphenol A bis(diphenyl phosphate) demonstrated moderate efficiency of these products in nylon 6 and nylon 6,6. It was shown that aromatic phosphates promote charring of the nylons, but the char yield is insufficient to rapidly extinguish the flame.

Introduction

There are numerous publications which show a variety of phosphorus-containing products under consideration in industrial and academic laboratories for use in nylons. In the introduction to this chapter we will overview the technical publications from the past 10 years on phosphorus-containing fire retardants in aliphatic nylons. This will be followed by a description of work on two phosphorus-containing compounds, resorcinol bis(diphenyl phosphate) and bisphenol A bis(diphenyl phosphate).

Red phosphorus.

Red phosphorus is particularly useful in glass-filled nylon 6,6 where the high processing temperature (about 280°C) excludes the use of most phosphorus-containing materials (*1*). The main disadvantages of red phosphorus as a fire retardant for nylons are its red color, flammability and the risk of generating highly toxic phosphine during storage or compounding (*2, 3*). In order to avoid oxidation of red phosphorus, it is usually stabilized with metal oxides and coated with a resin (*4, 5, 6, 7*). Red phosphorus shows synergistic action with highly charring polymers (*1, 8, 9*) and with some inorganic compounds (*10, 11, 12, 13, 14*). Vanadium oxides or metal vanadates (*15, 16, 17, 18*) are particularly efficient with red phosphorus. It is believed that vanadium phosphates are formed upon oxidation of red phosphorus and they are able to coordinate with amide groups and form layers of ordered char.

From the parallel trends of OI and the nitrous oxide (N_2O) index, it was experimentally shown that red phosphorus provides a condensed phase fire retardant action (*19*). The char yield produced from nylon 6 increases with an increasing content of red phosphorus. It was believed that red phosphorus oxidizes to phosphoric acid, which catalyzes char-formation. However, in a recent study we found (*20*) only ca. 4 % H_3PO_4 present in the char.

Kuper et al. (*21*) reported that red phosphorus mixed with nylon 6,6 had been oxidized even under nitrogen. By infrared and ^{31}P solid state NMR we detected (*12, 19*) phosphoric esters upon thermal decomposition of nylon 6 with red phosphorus in an inert atmosphere. This data, and the general observation that red phosphorus is particularly active in oxygen- or nitrogen-containing polymers led to an assumption on the possible direct interaction between phosphorus and the polymer at high temperature, possibly by free radical mechanism (19).

Ammonium polyphosphate and melamine phosphates

In our recent studies we found that ammonium polyphosphate (APP) can be incorporated into nylons 6 (*22*), 11 and 12 (*23*), 6,10 (*24*), or even into high melting nylon 6,6 (*24*) in spite of relatively low thermal stability of APP. Ammonium polyphosphate produces a continuous intumescent char on the surface which efficiently decreases the heat transfer to the polymer surface and slows the decomposition of nylon (*25*). Chemical processes for the formation of intumescent char were recently reviewed (*26, 27, 28*).

Some inorganic salts and metal oxides (*29, 30, 31, 32, 33*) are efficient co-additives with APP at relatively low loading, 1.5 – 3.0 wt. %. In the presence of these co-additives a less voluminous, but more heat protective and mass transfer restrictive, intumescent char is formed (*30, 33*). Binary metal-ammonium phosphates, which might simulate products of interaction of APP with metal oxides (*34, 35*) were prepared (*36*) and tested in nylon 6 (*37*). They showed only moderate fire retardant activity.

The fire retardancy of melamine phosphates in aliphatic nylons was reviewed by Weil (*38, 39, 40*) therefore we will limit our discussion to the most recent developments. In the series of patents Du Pont researchers (*41,42, 43, 44, 45*) claim the use of melamine phosphate or melamine pyrophosphate in combination with a catalytic amount of phosphotungstic acid or silicotungstic acid to obtain a V-0 rating in glass-reinforced nylons. DSM has recently introduced melamine polyphosphate as a new fire retardant for glass-reinforced polyamide 6,6 (*46*). The intumescent mechanism of fire retardant action was attributed to this product. Melam polyphosphate was found to be efficient in glass-reinforced blends of aromatic polyamide with nylon 6 and nylon 6,6 (*47*).

The mode of fire retardant action of melamine and its salts has been reported (*39, 48*) to be complex, involving both gas phase and condensed phase mechanisms. Our recent study (*49, 50*) showed destabilization of nylon 6; decreasing melt viscosity and promotion of flowing and dripping are the most evident contributions of melamine in the fire retardancy of nylons. On the other hand, melamine phosphates seem to exhibit a rather different mode of action since the phosphoric acid component involves nylons in the charring, similar to the action of ammonium polyphosphate.

Phosphonitrides

The highly cross-linked iminophosphazene $(PN_2H)_m$, known as phospham, is a light colored and very thermally stable product. Phospham provides a fire retardant effect in nylon 4,6, which requires very high processing temperature (340°C) (*51*). Recently BASF found (*52*) a specific grade of phospham prepared at 780°C to be particularly efficient in nylons 6 and 6,6.

It was found (*53, 54 ,55*) that the efficiency of phospham is similar to that of red phosphorus if compared on the basis of the phosphorus content in the formulation. Mechanistic studies showed (*56, 57, 58, 59*) that phospham increases the solid residue by promoting charring of nylon 6. Infrared and ^{31}P solid state NMR gave an indication that phospham interacts with nylon 6 and forms phosphoramides and phosphorimides which decompose at high temperature, producing char.

Phosphorus oxynitride $(PON)_n$ is another phosphorus-nitrogen containing product, which has potential in aliphatic nylons (*58, 60*). $(PON)_n$ increases the OI probably because it promotes extensive charring of nylon 6. In combination with Fe_2O_3 it provides a V-0 rating in the UL94 test. Another mode of the fire retardant action of $(PON)_n$ was believed (*61*) to be its tendency to create a low melting glass on the polymer surface.

Organo-Phosphorus Products

Trivalent phosphorus. Products with trivalent phosphorus easily undergo oxidation at high temperature, therefore there is a significant limitation on their use in thermoplastics. Nevertheless in the Atochem laboratories it was found (62) that some amine phosphites in combination with pentaerythritol provide a V-0 rating in nylon 11, which melts at relatively low temperature (190°C). Russian researchers (63) studied the fire retardant effect of triphenylphosphine in nylon 6. To improve the thermal-oxidative stability of triphenylphosphine, it was intercalated into montmorillonite type clay and this helped to further increase the oxygen index.

Phosphine oxides. In contrast to phosphites, phosphine oxides are among the most thermally and oxidatively resistant organophosphorus products. However, the relatively high cost of phosphine oxides prevents extensive development of fire retardant aliphatic nylons based on this class of organophosphates. This explains the lack of publications. The most recent US patent was granted to BASF in 1989 on use of triphenylphophine oxide in nylon 6/PPO blend (64).

Phosphinates. Glass fiber reinforced nylon 6,6 can be fire retarded by the addition of special phosphinic esters made by phosphorylation of bisphenol F or p-cresol novolac (65, 66, 67). In the series of patents to Penwalt (68, 69), Hoechst (70), Clariant (71, 72), Ticona (73, 74), and DSM (75), the use of Zn, Mn(II), Al, Mg, Ca or Sr salts of various phosphonic acids as primary fire retardant additives to aliphatic nylons was disclosed.

An adduct produced by reaction of 9,10-dihydro-9-oxa-10-phosphaphenanthrene 10-oxide with itaconic acid was suggested (76) for use as a monomer for nylons in fire retardant fibers. 2-Carboxyethyl(phenyl) phosphinic acid was also disclosed (77, 78) as a phosphorus-containing monomer for nylons.

Phosphonates. A mixture of cyclic phosphonates was suggested for use in aliphatic nylons (79, 80, 81). It was shown that 13% wt. of the additive provides a V-0 rating in neat nylon 6, whereas 15 - 20% wt. is needed for a V-0 rating in the glass reinforced system. In the Russian study (82) the effect of methylphosphonic acid or its diamide added to monomeric ε-caprolactam was investigated. They found that the maximum accumulation of phosphorus was observed with the combination of the diamide of methylphosphonic acid and AlO(OH). It was also reported that the diamide promotes graphitization of the char, which improves its thermal oxidative resistance.

Phosphates. Bis(bisphenol A-dicresyl) phosphate was found efficient in nylon 6/PPO blends at 25% wt. loading (83). The use of triphenyl phosphate in combination with melamine or melamine cyanurate helps to achieve a V-0 in the same blend at much lower loading (total 10% wt.) (64).

This paper shows the results of our recent project aimed at evaluation of the fire retardant efficiency of aromatic phosphates in two commercial nylons (nylon 6 and nylon 6,6).

Experimental

Materials

Nylon 6 and nylon 6,6 both unfilled and 33% glass-filled (GF) resins were provided by Du Pont. Phosphorus containing fire retardant additives used in this study were resorcinol bis(diphenyl phosphate) (RDP) and bisphenol A bis(diphenyl phosphate) (BDP), both from Akzo Nobel Functional Chemicals. Polyphenylene oxide (PPO) from GE, polyphenylene sulfide (PPS) from Aldrich, melamine, melamine cyanurate (MeCy) and melamine pyrophosphate (MePP) all from DSM were used as co-additives.

Compounding

Before compounding nylons were dried in a vacuum oven at 110°C for 2 hours. Formulations were compounded in a bowl mixer at 230°C for nylon 6 and 265°C for nylon 6,6 and then compression molded at the same temperatures. Specimens for combustion tests were cut from the molded plates.

Combustion.

Fire retardant performance of nylons was assessed either by the Oxygen Index (OI) test, according to the ASTM D 2863 standard, or by the UL-94 vertical combustion test. Specimens of 3x6x100 mm were used in the OI test and specimens of 1.6x12.5x125 mm in the UL-94 test.

Thermal Decomposition.

Thermal stability and charring performance of the fire retardant nylons were studied by thermogravimetry in argon or air atmosphere at a linear heating rate of 10°/min. Solid residues produced at different steps of thermal decomposition were analyzed by infrared spectroscopy.

Results and Discussion

Glass-filled nylons are more flammable than unreinforced polymers. The difference is most evident in the case of nylon 6,6, which has an OI of 26, whereas glass-filled nylon 6,6 has an OI of 22. This effect is mostly attributed to the tendency of neat nylons to drip away from the burning surface (*84, 85*). The glass-filler holds the melt on the surface and provides more surface exposed to the fire because of the "wicking effect".

Addition of RDP or BDP to nylon 6 does not affect the UL-94 rating but leads paradoxically to a decrease in oxygen index. Although neat nylon 6 shows a clean surface during combustion, some char appears on the surface of the samples containing RDP. The amount of char is probably enough to retard dripping as with glass-fiber reinforcement, but it is not sufficient to stop combustion. In contrast to nylon 6, both BDP and RDP tend to marginally increase the OI in unreinforced nylon 6,6.

Both BDP and RDP help to improve the fire retardancy of glass-filled nylons since the UL-94 rating goes from NC (non-classified) to V-2 when the additive is present.

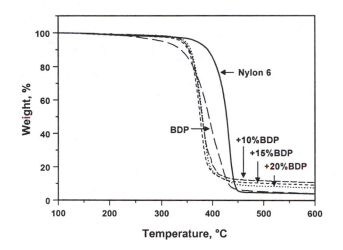

Figure 1. Thermogravimetry curves for nylon 6, BDP and nylon 6 added with different amounts of BDP. Heating rate 10 °C/min, argon flow 60 cm³/min.

Figures 1 and 2 show thermogravimetry data for nylon 6 and nylon 6,6 respectively, fire retarded by BDP. The curves with RDP are quite similar and are not shown here. The fire retardant additives destabilize both nylons, since the onset of weight loss occurs at 330-350°C instead of 370-380°C typical of these nylons. The phosphates help to increase the solid residue in both nylon 6 and nylon 6,6, however, surprisingly, the amount of residue decreases with an increase in loading.

The thermogravimetry curves of nylon 6,6 containing BDP (Fig. 2) are different from those of nylon 6 with BDP (Fig. 1). A two step process is observed in the case of nylon 6,6, which is indicative of a chemical interaction between the nylon and the phosphate.

The infrared spectra of the solid residues collected during the thermal decomposition of nylon 6,6 + 20% RDP are shown in Figure 3. The most

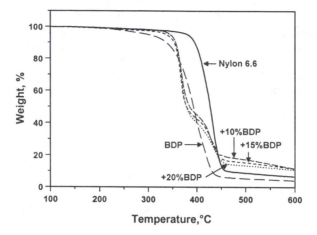

Figure 2. *Thermogravimetry curves for nylon 6,6, BDP and nylon 6,6 added with different amounts of BDP. Heating rate 10 °C/min, argon flow 60 cm³/min.*

intense absorption of RDP is seen at 966 cm⁻¹ in the initial formulation. This band has dropped significantly in intensity at 20% weight loss and almost disappears at 60% weight loss; this does not mean that RDP is lost. New bands appearing at 1025 and 1093 cm⁻¹ (20% weight loss) and 1241 cm⁻¹ (60% weight loss) are, probably, indicative of phosphorus containing products formed by the interaction of nylon 6,6 with RDP. They can be attributed either to phosphates or phosphoramides. We can not distinguish between these because absorptions of both P-O and P-N bonds lie in the same range 1050-950 cm⁻¹. Upon further heating to 70% and 85% of weight loss, the phosphorus containing structure loses its characteristic pattern; strong absorption at 1094 cm⁻¹ is probably indicative of some phosphorus retained in the residue.

Nylon 6,6 undergoes decomposition by a pathway similar to that of nylon 6, through scission of alkyl-amide bonds (Scheme 1, route "b") (*86*). However, some portion of nylon 6,6 also undergoes scission of CH_2-C(O) bonds (Scheme 1, route "a") (*24*), which is unusual for aliphatic nylons. The newly formed isocyanurate chain ends easily dimerize with the evolution of CO_2 and the production of carbodiimide. Since decomposition via route "a" is the only essential difference between thermal decomposition of nylon 6 and nylon 6,6, we may assume that phosphoric acidic moieties formed upon hydrolysis and thermal decomposition of aromatic phosphates will react with carbodiimide (*87*). This type of product is known (*88*) to cross-link and increase charring.

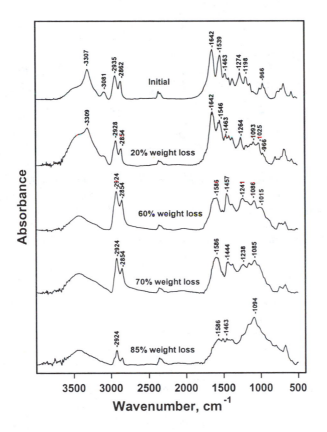

Figure 3. *Infrared spectra of nylon 6,6 containing RDP and solid residues obtained from this mixture at different steps of thermal decomposition. KBr pellets.*

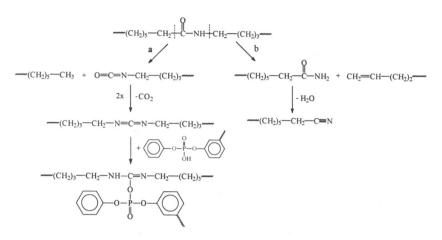

Scheme 1. Thermal decomposition of nylon 6,6 in the presence of aromatic phosphates.

As was shown above, RDP or BDP promote charring of nylon 6,6 but the amount of char is not enough to extinguish the flame, therefore we attempted to increase the amount of char by using co-additives (Table 1).

Table 1. Combustion performance of nylon 6,6 fire retarded by RDP or BDP with co-additives

No.	Additive		Co-additive		UL-94	OI
	Name	Wt. %	Name	Wt. %		
1	RDP	15	-	-	NC	28
2	-"-	10	PPS	5	V-2	28
3	-"-	10	PPO	10	NC	27
4	-"-	10	MePy	5	V-2	29
5	-"-	10	MeCy	5	V-2	27
6	BDP	20	-	-	V-2	27
7	-"-	10	PPS	10	V-2	30
8	-"-	10	PPO	10	V-2	28
9	-"-	10	Melamine	10	V-2	29

The co-addition of highly charring polymers (polyphenylene oxide or polyphenylene sulfide) helps to improve the fire retardancy in the case of BDP, but the fire retardancy deteriorates in the case of RDP. Nitrogen-rich products, such as melamine cyanurate or melamine phosphate, were found to be moderately synergistic with RDP and BDP in aliphatic nylons.

Conclusions

In spite of the variety of phosphorus-containing products in the technical literature, only red phosphorus and melamine phosphates seem to find commercial use. The search for halogen-free fire retardants, providing good balance of fire retardancy and physical properties in aliphatic nylons, continues.

The fire retardant action of resorcinol bis(diphenyl phosphate), RDP, and bisphenol A bis(diphenyl phosphate), BDP, was studied in nylon 6 and 6,6. Both RDP and BDP do not affect the flammability of unreinforced nylons as measured by UL-94. In the glass-filled nylons, the presence of RDP and BDP enable a V-2 rating in the UL-94 test. Some highly charring polymers and nitrogen-containing additives improve fire retardancy efficiency.

As shown by thermogravimetry, the aromatic phosphates promote charring of the nylons, however the amount of the char formed is, probably, not enough to efficiently extinguish the polymers. In fact, the infrared spectra of solid residues shows that RDP is involved in the charring of nylon 6,6. It is assumed that acidic moieties produced upon hydrolysis and thermal decomposition of aromatic phosphates react with carbodiimide functionalities formed at the thermal decomposition of nylon 6,6.

References

1 Huggard, M.T. In *Pros. Conf. Recent Adv. Flame Retardancy Polym. Mater.*, Stamford, CT, 1995.
2 Bordero, S. In *Proc. Conf. Recent Adv. Flame Retardancy Polym. Mater.*, Stamford, CT, 1993.
3 Weil, E.D. In *Proc. Conf. Recent Adv. Flame Retardancy Polym. Mater.*, Stamford, CT, 2000.
4 Bonin, Y.; LeBlanc, J. to Rhone-Poulenc, US Patent 4921896, 1990.
5 Bonin Y.; LeBlanc, J. to Rhone-Poulenc, US Patent 4985485, 1991.
6 Steiert, P.; Weiss, H.-P.; Plachetta, C.; Baierweck, P.; Muehlbach, K.; Gareiss, B. to BASF, US Patent 5049599, 1991.
7 Steiert, P.; Weiss, H.-P.; Plachetta, C.; Baierweck, P.; Muehlbach, K.; Gareiss, B. to BASF, US Patent 5135971, 1992.
8 Huggard, M.T. In *Proc. Conf. Recent Adv. Flame Retardancy Polym. Mater.*, Stamford, CT, 1992.
9 Muehlbach, K.; Steiert, P.; Vogel, W.; Kurps A. to BASF, US Patent 5260359, 1993.
10 Gareiss, B.; Klatt, M.; Gorrissen H. to BASF, PCT Patent Application WO98/23676, 1998.
11 Bonin, Y.; LeBlanc, J. to Rhone-Poulenc, US Patent 5466741, 1995.

224

12 Levchik, G.F.; Levchik S.V.; Camino, G.; Weil, E.D. In *Fire Retardancy of Polymers. The Use of Intumescence*, Le Bras M.; Camino, G.; Bourbigot, S.; Delobel, R., Eds. The Royal Soc. Chem. Publ.: London, 1998, pp 304-315.

13 Moriwaki, T. to Kishimoto Sangyo, European Patent Application, EP 0592942, 1998.

14 Klatt, M.; Grutke, S.; Heitz, T.; Rauschenberger, V.; Plesnivy, T.; Wolf, P.; Wuensch, J.; Fischer, M. to BASF, PCT Patent Application WO 98/36022, 1998.

15 Kodolov, V.I.; Knyazeva, L.F.; Hramavaya, G.S.; Mihailov, V.I.; Povstugar, V.I.; Kibenko V.D. *Plast Massy* **1985** (11), 56-57.

16 Kodolov V.I.; Mikhailov V.I.; Larionov K.I.; Tyurin S.A. In *Proc. 2nd Beijing Int. Symp./Exhibit. Flame Retardants*, Beijing, 1993, pp 381-386.

17 Bulgakov, V.K.; Kodolov, V.I.; Lipanov, A.M. *Modeling of Combustion of Polymeric Materials,* Khimiya: Moscow, 1990, p 205 (in Russian).

18 Kibenko V.D.; Kodolov V.I.; Dorfman A.M.; Elanskii A.V.; Vasiljev V.A. In *Proc. 2nd Beijing Int. Symp./Exhibit. Flame Retardants*, Beijing, 1993, pp 33-37.

19 Levchik, S.V.; Levchik, G.F.; Balabanovich, A.I.; Camino, G.; Costa, L. *Polym. Degrad. Stab.* **1996**, *54*, 217-222.

20 Levchik, G.F.; Vorobyova, S.A.; Gorbarenko, V.V.; Levchik, S.V. *J. Fire Sci.,* **2000**, *18*, 172-182.

21 Kuper, G.; Hormes, J. Sommer, K. *Macromol. Chem. Phys.* **1994**, *195*, 1741-1753.

22 Levchik, S.V.; Costa, L.; Camino, G. *Polym. Degrad. Stab.* **1992**, *36*, 229-237.

23 Levchik, S.V.; Costa, L.; Camino, G. *Polym. Degrad. Stab.* **1994**, *43*, 43-54.

24 Levchik, S.V.; Costa, L.; Camino, G., *Polym. Degrad. Stab.* **1992**, *36*, 31-42.

25 Levchik, S.V.; Camino, G.; Costa, L.; Levchik, G.F. *Fire Mater.* **1995**, *19*, 1-10.

26 Levchik, S.V. *Vestn. Belar. Univers. Ser. Khim. Biolog. Geogr.* **1998**, *(1)*, 3-13.

27 Lesnikovich, A.I.; Bogdanova, V.V.; Levchik, S.V.; Levchik, G.F., In *Chemical Problems of Development of New Materials and Technologies*, Sviridov, V.V. Ed., Belgosuniversitet: Minsk, 1998, pp 145-168.

28 Levchik, S.V.; Weil, E.D.; Lewin, M., *Polym. Int.,* **1999**, *48*, 532-557.

29 Levchik, S.V.; Levchik, G.F.; Selevich, A.F.; Camino, G.; Costa, L., In *Proc. 2nd Beijing Int. Symp./Exibit. Flame Retardants*, Beijing, 1993, pp 197-202.

30 Levchik, S.V.; Levchik, G.F.; Camino, G.; Costa, L. *J. Fire Sci.* **1995**, *13*, 43-58.

31 Levchik, G.F.; Levchik, S.V.; Lesnikovich, A.I. *Polym. Degrad. Stab.* **1996**, *54*, 361-363.

32 Levchik, G.F.; Levchik, S.V.; Lesnikovich, A.I. *Dokl. AN Belarusi* **1996**, *40* (5) 74-77.

33 Levchik, S.V.; Levchik, G.F.; Camino, G.; Costa, L.; Lesnikovich, A.I. *Fire Mater.* **1996**, *20*, 183-190.
34 Levchik, G.F.; Selevich, A.F.; Levchik, S.V.; Lesnikovich, A.I. *Thermochim. Acta* **1994**, *239*, 41-49.
35 Levchik, G.F.; Levchik, S.V.; Sachok, P.D.; Selevich, A.F.; Lyakhov, A.S.; Lesnikovich, A.I. *Thermochim. Acta*, **1995**, *257*, 117-125.
36 Selevich, A.F.; Levchik, S.V.; Lyakhov, A.S.; Levchik, G.F.; Catala, J.-M. *J. Solid State Chem.* **1996**, *125*, 43-46.
37 Levchik, G.F.; Levchik, S.V.; Selevich, A.F; Lesnikovich, A.I.; Lutsko, A.V. In *Fire Retardancy of Polymers. The Use of Intumescence*, Le Bras M.; Camino, G.; Bourbigot, S.; Delobel, R., Eds. The Royal Soc. Chem. Publ.: London, 1998 p.280-289.
38 Weil, E.D. In *Proc. Conf. Recent Adv. Flame Retardancy Polym. Mater.*, Stamford, CT, 1994.
39 Weil, E.D.; Choudhary, V. *J. Fire Sci.* **1995**, *13*, 104-126.
40 Weil, E.D. *Plast. Compound.* **1994**, May/June, 31-39.
41 Martens, M.M.; Kasowski, R.V. to Du Pont, US Patent 5618865, 1997.
42 Martens, M.M.; Kasowski, R.V.; Cosstick, K.B.; Penn, R.E. to Du Pont, US Patent 5708065, 1998.
43 Kasowski, R.V. to Du Pont, US Patent 5859099, 1999.
44 Kasowski, R.V.; Martens, M.M. to Du Pont, PCT Patent Application WO 98/45364, 1998.
45 Kasowski, R.V. to Du Pont, PCT Patent Application WO 99/46250, 1999.
46 Boelens, C.; Grabnex, R. In *Proc. Conf. Recent Andv. Flame Retardancy Polym. Mater.*, Stamford, CT, 1999.
47 Horino, M.; Watanabe, N. to Mitsubishi Engineering Plastics, European Patent Application EP 0881264, 1998.
48 Weil, E.D.; Zhu, W. In *Fire and Polymers II. Materials and Tests for Hazard Prevention*, Nelson, G.L. Ed.; ACS Symposium Series: Washington, DC, Vol. 599, 1995, pp 199-216.
49 Levchik, S.V.; Balabanovich, A.I.; Levchik, G.F.; Costa, L. In *Proc. Conf. Recent Adv. Flame Retardancy Polym. Mater.* Stamford, CT 1996.
50 Levchik, S.V.; Balabanovich, A.I.; Levchik, G.F.; Costa, L. *Fire Mater.* **1997**, *21*, 75-83.
51 Weil, E.D.; Patel, N.G. to Stamicarbon, US Patent 4946885, 1990.
52 Gareiss, B.; Schneider, H.-M.; Weber, M. to BASF, German Patent, DE 19615230, 1997.
53 Weil, E.D.; Patel, N.; Huang, C.-H. In *Proc. Conf. Recent Adv. Flame Retardancy Polym. Mater.* Stamford, CT, 1993.
54 Weil, E.D.; Patel, N.; Huang, C.-H.; Zhu, W. In *Proc. 2nd Beijing Int. Symp./Exhibit. Flame Retardants*, Beijing, 1993, pp 285-290.
55 Weil, E.D.; Patel, N.G. *Fire Mater.* **1994**, *18*, 1-7.

226

56 Costa, L.; Catala J.-M.; Gibov, K.M.; Levchik, S.V.; Khalturinskii, N.A. In *Fire Retardancy of Polymers. The Use of Intumescence*, Le Bras, M.; Camino, G.; Bourbigot. S.; Delobel, R., Eds.; The Royal Soc. of Chem. Publ.: London, 1998, p.76-87.

57 Levchik, G.F.; Levchik, S.V.; Lesnikovich, A.I. *Dokl. AN Belarusi*, **1999**, *43* (1), 53-56.

58 Balabanovich, A.I.; Schnabel, W.; Levchik, G.F.; Levchik, S.V.; Wilkie, C.A. In *Fire Retardancy of Polymers. The Use of Intumescence*, Le Bras, M.; Camino, G.; Bourbigot. S.; Delobel, R., Eds.; The Royal Soc. of Chem. Publ.: London, 1998, pp 236-251.

59 Balabanovich, A.I.; Levchik, S.V.; Levchik, G.F.; Schnabel, W.; Wilkie, C.A. *Polym. Degrad. Stab.*, **1999**, *64*, 191-195.

60 Levchik, S.V.; Levchik, G.F.; Balabanovich, A.I.; Weil E.D.; Klatt, M. *Angew. Makromol. Chem.*, **1999**, *264*, 48-55.

61 Weil, E.D. In *Improved Fire- and Smoke-Resistant Materials for Commercial Aircraft Interiors*, National Academy Press: Washington, DC, 1995, pp 129-150.

62 Poisson, P.; Rivas, N.; Deloy, P. to Atochem, US Patent 4879327, 1989.

63 Lomakin, S.M.; Usachev, S.V.; Koverzanova, E.V.; Ruban, L.V.; Kaligina, I.G.; Zaikov, G.E. In *Proc. Conf. Recent Adv. Flame Retardancy Polym. Mater.* Stamford, CT, 1999.

64 Taubitz, C.; Gausepohl, H.; Rochlke, K. to BASF, US Patent 4866114, 1989.

65 Fuhr, K.; Muller, F.; Ott, K.-H.; Al-Sayed, A.; Muller, P.-R.; Wanddel, M. to Bayer, US Patent 5021488, 1991.

66 Fuhr, K.; Muller, F.; Ott. K.-H.; Al-Sayed, A.; Muller, P.-R.; Wanddel, M. to Bayer, US Patent 5102931, 1992.

67 Ostlinning, E.; El Sayed, A.; Sommer, K.; Frohlen, H.G. to Bayer, US Patent 5242960, 1993.

68 Sandler, S.R. to Pennwalt, US Patent 4208321, 1980.

69 Sandler, S.R. to Pennwalt, US Patent 4208322, 1980.

70 Kleiner, H.-J.; Budzinsky, W.; Kirsch, G. to Hoechst, European Patent Application, EP 0794191, 1997.

71 Jenewein, E.; Nass, B. to Clariant, European Patent Appl., EP 0899296, 1999.

72 Nass, B.; Wanzke, W. to Clariant, European Patent Appl., EP 0896023, 1999.

73 Kleiner, H.-J.; Budzinsky, W.; Kirsch, G.; to Ticona, US Patent 5773556, 1998.

74 Kleiner, H.-J.; Budzinsky, W. to Ticona, PCT Patent Application, WO 98/39381, 1998.

75 Hulskotte, R. to DSM, PCT Patent Application, WO 99/02606, 1999.

76 Rieckert, H.; Dietrich, J.; Keller, H. German Patent, DE 19711523, 1998.

77 Asrar, J. to Solutia, US Patent 5750603, 1998.

78 Asrar, J. to Solutia, PCT Patent Application WO 98/54248, 1998.

79 Green, J. In *Proc. Conf. Recent Adv. in Flame Retardancy Polym. Mater.* Stamford, CT, 1992.

80 Huggard, M.T. In *Proc. Conf. Recent Adv. Flame Retardancy Polym. Mater.* Stamford, CT, 1994.

81 Klatt, M.; Heitz, T.; Gareiss, B. to BASF, PCT Patent Application WO 98/11160, 1998.

82 Zubkova, N.S.; Tyuganova, M.A.; Mihailova, E.D.; Duderov, N.G. *Khim. Volokna* **1992** *(5),* 39-40.

83 Kakegawa, J.; Takayama, S. to Asahi, US Patent 5455292, 1995.

84 Gilleo, K.B. *Eng. Chem. Prod. Res. Develop.* **1974**, *13*, 139-143.

85 Wharton, R.K. *J. Appl. Polym. Sci.* **1982**, *27*, 3193-3197.

86 Levchik, S.V., Costa, L.; Camino, G. *Makromol. Chem., Macromol. Symp.* **1993**, *74*, 95-99.

87 Grassie, N.; Perdomo Mendoza, G.A. *Polym. Degrad. Stab.,* **1985**, *11*, 145-166.

88 Grassie, N.; Zulfiqar, M., In *Developments in Polymer Stabilization*, Scott, G. Ed.; Appl. Sci. Publ.: London, 1979, Vol.1, pp 1-43

Chapter 18

Zinc Borates: 30 Years of Successful Development as Multifunctional Fire Retardants

Kelvin K. Shen

U.S. Borax Inc., 26877 Tourney Road, Valencia, CA 91355

Zinc borates can function as flame retardant, smoke suppressant, afterglow suppressant, and anti-tracking agent in both halogen-containing and halogen-free polymers. This paper will review the development and recent advances in the use of these multifunctional fire retardants in electrical/electronic, transportation, and building product applications. The mode of action of these fire retardants in both halogen-containing and halogen-free polymers will also be reviewed.

Introduction

The use of an organic chlorine or bromine source to impart fire retardancy to polymers is well known in the plastics and rubber industries. To further enhance the fire test performance of halogen-containing polymers, antimony oxide is usually used as a synergist. In recent years, however, much effort has been expended to find either partial or complete substitutes for antimony oxide. This effort has been spurred by the desire to achieve better smoke suppression, better cost/performance balance, and by environmental concerns. One of the most commercially successful substitutes is zinc borate (1-7). Depending on the reaction conditions, a host of zinc borates with different mole ratios of $ZnO:B_2O_3:H_2O$ can be produced. In the 1970s, a unique form of zinc borate with a molecular formula of $2ZnO \cdot 3B_2O_3 \cdot 3.5H_2O$ was discovered (5). In contrast to previously known zinc borates, this material (known in the trade as *Firebrake*®ZB), is stable to 290°C (Figure 1). During the past thirty years, it was developed into a major commercial fire retardant used in both halogen-containing and halogen-free polymer systems. In recent years, due to the demand of high production throughput and thin-walled electrical parts, engineering plastics are being processed at increasingly higher temperatures. To meet the market demand of engineering plastics, U.S. Borax recently developed an anhydrous zinc borate, *Firebrake*®500 ($2ZnO \cdot 3B_2O_3$), that is stable to at least 450°C and $4ZnO \cdot B_2O_3 \cdot H_2O$, *Firebrake*® 415, that is stable to about 415°C. This paper will review recent advances in the use of zinc borates as multifunctional fire retardants in polymers. Special emphasis will be on electrical/electronic, transportation, and building material applications.

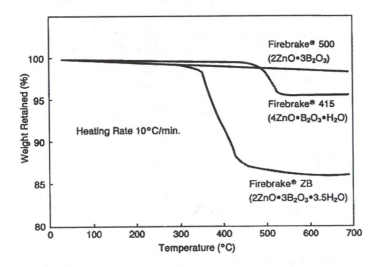

Figure 1. Thermogravimetric analysis of zinc borates.

Results and Discussion

Zinc borates are multifunctional fire retardants. They can provide the following benefits in fire retardant polymer systems.

- Synergist of most halogen sources – They are synergists of both chlorine and bromine-containing flame retardants. Their efficacy depends on the type of halogen source (aliphatic vs. aromatic) and the base polymer used.

- Synergist of antimony oxide - In certain polymers and in the presence of a certain halogen source, zinc borates can display synergistic effects with antimony oxide in fire retardancy.
- Partial or complete replacement of antimony oxide - Their efficacy depends on both the polymer used and the halogen source used.
- Smoke suppressant - In contrast to antimony oxide, zinc borates are predominately a condensed phase flame retardant and can function as smoke suppressants.
- Afterglow suppressant - The B_2O_3 moiety in zinc borates are responsible for its afterglow suppression effect.
- Fire retardancy synergism with alumina trihydrate (ATH) in many halogen-containing systems.
- Promoter of char formation and prevent dripping - The zinc oxide moiety of zinc borates can promote char formation with halogen sources; the B_2O_3 moiety can stabilize the char.
- Improvement of the Comparative Tracking Index (CTI) - The B_2O_3 moiety of zinc borates is believed to play a major role for this unique property.
- Formation of porous ceramic with ATH that act as an insulator in protecting the underlying polymers.

Some of the recently discovered benefits from U.S. Borax and the literature are as follows.
- Reduce rate of heat release
- Improve aged elongation properties of polyolefins
- Improve thermal stability of halogen/antimony oxide system
- Improve corrosion resistance by replacing antimony oxide
- Improve laser marking quality of polymers
- Function as flame retardants in certain halogen-free systems

Zinc borate ($2ZnO \cdot 3B_2O_3 \cdot 3.5H_2O$)

This zinc borate is of special commercial importance (Figure 2) (5). It is the most widely used among all of the known zinc borates; it has been used extensively in PVC, nylon, epoxy, polyolefin, and elastomers (1-4). It can function as a flame retardant, smoke suppressant, afterglow suppressant, and anti-tracking agent. For example, the use of this zinc borate as a smoke suppressant in flexible PVC is illustrated in Figure 3. In a recent Cone Calorimeter study in flexible PVC, partial replacement of antimony oxide with the zinc borate can not only reduce the rate of heat release, but also carbon monoxide production at a heat flux of 35 kW/m^2 (Figure 4) (8).

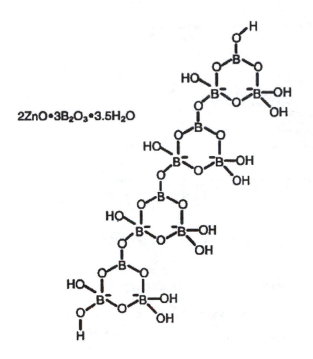

$2ZnO \cdot 3B_2O_3 \cdot 3.5H_2O$

Figure 2. *Proposed molecular structure of zinc borate.*

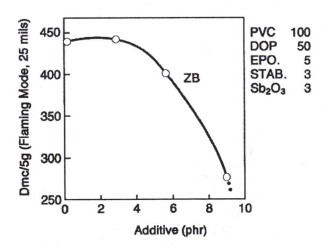

Figure 3. *Effect of zinc borate in ASTM E-662 smoke test of flexible PVC*

232

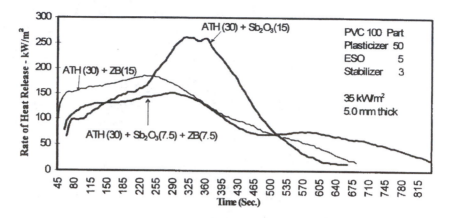

Figure 4. Effect of the zinc borate on rate of heat release in flexible PVC.

The use of the zinc borate as an afterglow suppressant in syndiotactic polystyrene, a new engineering plastic (9), is illustrated in Table I.

Table I. Ignition Resistance of Syndiotactic Polystyrene (S-PS)

Components	1	2	3	4
S-PS, Brominated FR				
Plus Other Additives	96.7	96.0	97.0	97.0
Sb_2O_3	3.3	4.0	1.5	2.0
Zinc Borate	-	-	1.5	1.0
Properties				
Uncolored Sample				
UL-94 (1.6 mm)	V-O	V-O	V-O	-
Avg. Afterglow Time (sec.)	8.6	9.0	9.4	-
Sample with Carbon Black				
UL-94 (1.6 mm)	V-1	V-1	V-O	V-O
Avg. Afterglow Time (sec.)	32.8	34.8	13.8	18

(Examples (wt.%))

In glass reinforced nylon 6,6 containing Dechlorane Plus (an alicyclic chlorine source), the zinc borate can replace antimony oxide completely. The zinc borate functions not only as a flame retardant synergist of Dechlorane Plus, but also as an anti-tracking agent (i.e. significant improvement in the Comparative Tracking Index) (10,11). With brominated polystyrene as the halogen-source, the zinc borate can replace antimony oxide completely but requires the use of a slightly higher level of the bromine source or synergist in order to maintain the UL-94 V-O rating. Most importantly, the use of zinc borate can increase the thermal stability of the fire retardant polymer system, as evidenced by the color stability and weight retention at 340°C (12,13) (Table II).

Table II. Flame Retardant Nylon 6,6

	Examples (wt.%)	
Components	1	2
Nylon 6,6	45.1	45.0
Glass Fiber	25.0	25.0
Zinc Borate	-	5.0
Bromine Compound	23.5	23.5
Antimony Compound	4.9	-
Surlyn	1.5	1.5
Properties		
TGA Retention (%) at 340 °C for 30 min.	63.0	95.0
Tensile Strength (psi)	27.0	26.7
Elongation (%)	2.8	2.8
UL-94 (1/32 in.)	V-O	V-O

Martens (14) et al. reported the use of zinc borate in halogen-free nylon 6,6 for a dramatic improvement in CTI (Table III).

Table III. Halogen-Free, Flame Retardant Nylon 6,6

	Examples (wt. %)			
Components	1	2	3	4
Nylon 6,6	44.9	44.9	44.9	44.9
Fiber Glass	25.1	25.1	25.1	25.1
Melamine Pyrophosphate	30	25	20	25
Melamine Cyanurate	-	5	10	-
Zinc Borate	-	-	-	5
Properties				
UL-94 (0.16 mm)	V-O	V-O	V-1	V-0
CTI (Volts)	275	400	350	600

Interestingly enough, when used in conjunction with a hindered phenol stabilizer in polyolefins containing decabromodiphenyl oxide and antimony oxide, the zinc borate was reported to impart a remarkable retention of aged elongation property at 180°C (Table IV) (15). This improvement plays an important role in polyolefin wire and cable application.

Table IV. Flame Retardant Polyolefins

Components	Examples (pbw) 1	2	3
Polyethylene (low density)	50	50	-
Ethylene-Vinyl Acetate	50	50	50
Ethylene-Alkyl Acrylate	-	-	50
Irgonox 1010	2	2	2
Zinc Borate	10	-	10
Sb_2O_3	5	5	5
Decabromodiphenyl Oxide	20	20	20
Properties			
Initial Tensile Strength (Kg/mm^2)	1.5	1.6	1.3
Initial Elongation (%)	400	420	350
180°C, 7 Days Tensile Strength (Kg/mm^2)	1.7	0.8	1.3
180°C, 7 Days Elongation (%)	380	10	340
Modulus (Kg/mm^2)	9.8	8.8	7.5
60°C, 30 Days (Kg/mm^2)	10.1	8.7	7.4
Volume Resistivity (Ω cm)	4×10^{16}	4×10^{16}	4×10^{16}

Recently, it was reported that the zinc borate can inhibit the stainless steel corrosion from PET containing an aromatic bromine source and antimony oxide at 320°C (16) (Table V).

In halogen-free ethylene-vinyl acetate (EVA), partial replacement of ATH with the zinc borate can result in a porous ceramic-like residue during polymer combustion. This ceramic residue can prevent short-circuiting and protect underlying polymer from further combustion. The detailed results were reported at the 1990 Fire and Polymer Symposium (3,17). Recent studies at the University of Lille demonstrated that partial replacement of ATH with the zinc borate can also decrease Rate of Heat Release, increase Time to Ignition, and delay the degradation of the protective surface structure (18).

The use of zinc borate in conjunction with red phosphorus and ATH in polyolefin was also reported. Apparent synergy was observed for this combination as evidenced by Oxygen Index test results (19). In the semiconductor encapsulation application, the use of zinc borate as a sole fire retardant in silica filled epoxy to meet UL-94 V-O was also reported (20). The synergy between zinc borate and ammonium polyphosphate was claimed in several applications (21).

Table V. Flame Resistant Polyalkylene Terephthalate
Examples (pbw)

Components	1	2	3	4	5
PET	100	100	100	100	100
Brominated Bisphenol A-Epoxy	15	15	15	15	15
Sb_2O_3	2	2	2	2	2
Zinc Borate	-	-	0.8	-	0.4
Zinc Oxide	-	0.8	-	-	0.4
Zinc Sub. Hydrotalcite	-	-	-	1.0	1.0
Properties					
Corrosion Test (300°C)					
40 min	Poor	V. Poor	Fair	Fair	Good
60 min	V. Poor	-	Poor	Fair	Good
80 min	-	-	V. Poor	Poor	Good

Zinc borate ($2ZnO \cdot 3B_2O_3$)

This anhydrous zinc borate is recommended for use in engineering plastics processed at above 300°C which is the upper limit of the zinc borate ($2ZnO \cdot 3B_2O_3 \cdot 3.5H_2O$). It is reported to be an effective smoke suppressant in fluoropolymers for plenum cable applications (22). Recently it was claimed that this anhydrous zinc borate can replace antimony oxide completely in high temperature nylon applications (13). For aircraft application, this material is also reported to be effective in reducing rate of heat release in polyether ketones and polysulfones (23) (Table VI). It has also been claimed to enhance laser marking quality of fire retardant engineering plastics (24).

Table VI. Poly(biphenyl ether sulfone) and Poly(aryl ether ketone) Blend
Examples (parts by wt.)

Composition	1	2	3	4
Poly(biphenyl ether sulfone)/ Poly(aryl ether ketone) 65/35	100	100	100	100
Zinc Borate ($2ZnO \cdot 3B_2O_3$)	-	4	8	8
Polytetrafluoroethylene	-	-	-	2
Properties				
Max. Heat Release Rate (KW/m^2)	91	54	44	35
Impact Strength (ft.-lb.)	-	126	106	95

Zinc borate (4ZnO·B₂O₃·H₂O)

This newly developed zinc borate (25) has an unusually high onset of dehydration temperature of >415°C. It is recommended for use in engineering plastics as a synergist of halogen sources. It can also be used as a smoke suppressant in flexible PVC. Interestingly enough, a recent study showed that partial replacement of magnesium hydroxide with this zinc borate in halogen-free EVA results in the formation of a rigid residue. Cone calorimeter test showed that the zinc borate can also decrease Rate of Heat Release and delay the second heat release peak (26).

Mode of action

Zinc borate in halogen-containing systems such as flexible PVC, is known to markedly increase the amount of char formed during polymer combustion; whereas the addition of antimony oxide, a vapor phase flame retardant, has little effect on char formation (1). Zinc borate can react with hydrogen halide, released from thermal degradation of a halogen source, to form zinc chloride, zinc hydroxychloride, boric oxide, water and possibly a small amount of boron trichloride. The zinc species remaining in the condensed phase can alter the pyrolysis chemistry by catalyzing dehydro-halogenation and promoting cross-linking between polymer chains, resulting in increased char formation and decrease in both smoke production and flaming combustion. The released boric oxide, a low melting glass, can stabilize the char and inhibit afterglow through glass formation. The water released between 290-450°C (accompanied by an endothermicity of 500 Joules/gram) can promote the formation of a foamy char. The mode of action between zinc borate and an aromatic halogen source is, however, not clearly understood.

In halogen-free systems, previous study showed that partial replacement of ATH with zinc borate ($2ZnO·3B_2O_3·3.5H_2O$) in EVA can favorably alter the oxidative-pyrolysis chemistry of the base polymer, as shown by the decrease in the exothermicity in the oxidative pyrolysis range and the delay of the peak oxidative-pyrolysis weight loss rate (17). The combination of ATH and zinc borate can form a porous and ceramic-like residue at temperatures above 550°C (probably in the range of 700-800°C). This residue is an important thermal insulator for the unburned, underlying polymer (Figure 5). Figure 6 shows the use of zinc borate alone in EVA can result in the formation of a glassy residue. It should be noted that, due to the flaming combustion at the early stage of the air pyrolysis, the sample is actually being exposed to temperature much higher than 550°C.

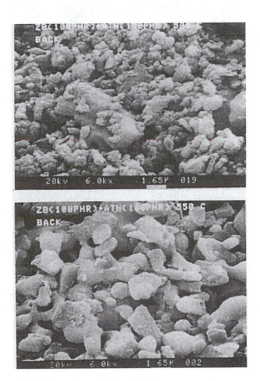

Figure 5. *Scanning electron micrographs (6000X) of residues of air-pyrolyses of EVA containing the zinc borate ($2ZnO \cdot 3B_2O_3 \cdot 3.5H_2O$) (33%) and ATH (33%). At 500°C (top picture), a powdery residue was obtained; no sintering took place between the zinc borate (small particles) and ATH (large particles). At 550°C (bottom picture), a hard and porous residue was obtained. The zinc borate acted as a sintering aid.*

238

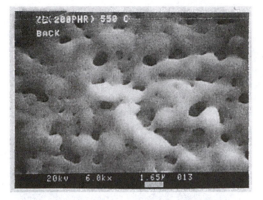

Figure 6. *Scanning electron micrographs (6000X) of residues of air-pyrolysis of EVA containing the zinc borate $(2ZnO \cdot 3B_2O_3 \cdot 3.5H_2O)$ (50%) but no ATH.*

Conclusions

- Zinc borates are multifunctional fire retardants. They can function as a flame retardant, smoke suppressant, afterglow suppressant, and anti-tracking agent.
- They can display synergistic/beneficial effect with other synergists or additives such as antimony oxide, ATH, magnesium hydroxide, red phosphorus, and ammonium polyphosphate in both halogen-containing and halogen-free polymers.
- The choice of optimal synergistic combination depends on the base polymer and halogen source used (i.e. additive vs. reactive approach, aromatic vs. aliphatic halogen source).
- Zinc borates alone can function as fire retardants in some polymers.

References

1. Shen, K.K. *Plastics Compounding*, **1985**, *8(5)*, 66-78.
2. Shen, K.K.; O'Connor, R. in *"Plastics Additives"*, Pritchard, G., Ed.; Chapman & Hall, London, 1998, pp 268-286.
3. Shen, K.K.; Griffin, T.S. in *"Fire and Polymers"*-ACS Symposium Series 425; Nelson, G.L., Ed.; American Chemical Society, 1990, pp. 157-177.
4. Shen, K.K.; Schultz, D, *"Rubber Processing and Performance,"* Hanser Publication , **2000** (in press).
5. Nies, N.; Hulbert P. *U.S. Patent* 3,549,316 and 3,649,172.
6. Shen, K.K.; Ferm D.J. *Sixth BCC Conference on Fire Retardancy*, Stamford, CT; **1995**.
7. Ferm, D.J.; Shen K.K. *Proceedings International Conference on Fire Safety*, San Francisco, **1992**.

8. Ferm, D.J.; Shen, K.K. *Plastics Compounding*, **1994**, *17(6)*40-44.
9. *Research Disclosure*, **June 1998**/737. (RD 0410059).
10. Shen, K.K.; Ferm D.J. *Eighth BCC Conference on Fire Retardancy*, Stamford, Conn., **1997**.
11. Markezich, R.; Ilardo, C. *SPE ANTEC Proceedings*, **1988**, 1412.
12. Miyabo, A.; Koshida, R. *PCT Int. Appl. WO* 9,814,510, **1998**.
13. Martens, M.M; Koshido, R; Tobin, W; Willis, J. *PCT Int. Appl. WO 9,947,597*, **1999**.
14. Martens, M.M; Kasowski, R; Cosstick, K.B; Penn, R.E. *PCT Int. Appl. WO* 9,723,565, **1997**.
15. Yadoshima, S *Japan Kokai Tokkyo Koho* JP 03,259,938, **1991**.
16. Japan Kokai Tokkyo Koho JP 11,189,711-A, **1999**.
17. Shen, K.K. *Plastics Compounding*, **1988**, *11(7)*, 26-32.
18. Bourbigot, S; LeBras, M.; Leeuwendal, R.M; Shen, K.K.; Schubert, D.M. *Polym. Degrad. Stab.*, **1999**, *64*, 419-425.
19. *Japan Kokai Tokkyo Koho* JP 58,109,546, **1983**.
20. Kamikooryma, *Japan Kokai Tokkyo Koho* JP 06,107,914, **1994**.
21. Istvan, B.; Marosi, G.; Peter, A.; Istvan.; Andras, T.; M. Maatoug.; Karoly, S. Muanyag Gumi **1997**, *34*, 237-243.
22. Chandrasekaran, S.; Kundel, N.K.; Garg, B.; Chin, H.B. *U.S. Patent* 4,957,961 **1990**.
23. Kelley, W.; Matzner, M.; Patel, S. *PCT Int. Appl. WO* 91 15,539, **1991**.
24. Kato, H. *European Patent Application* 0 675 001 A1 **1995**.
25. Schubert, D. *U.S. Patent* 5,342,553 and 5,472,644.
26. Carpentier, F.; Bourbigot, S.; LeBras, M.; Delobel, R. private communication.

Chapter 19

Oxygenated Hydrocarbon Compounds as Flame Retardants for Polyester Fabric

Kobie Bisschoff and W. W. Focke*

Institute of Applied Materials, Department of Chemical Engineering,
University of Pretoria, Lynnwood Road, Pretoria, South Africa, 0002
(email: xyris@mweb.co.za)

Selected oxygenated hydrocarbons impart a measure of flame resistance to unsized polyester fabric. The postulated mechanism involves a combination of enhanced melt dripping and fiber shrinkage effects. These enhance heat removal from the fabric front exposed to the flame making it more difficult to ignite the sample.

Introduction

The purpose of flame retardant treatments is to reduce the rates of burning and flame spreading. With flame retarded fabrics it is usually required that they must pass some type of standard vertical flame test (e.g. NCB 245, UL 94V, NFPA 701, BS 5867, DIN 4102, X65020-1991, etc.). Conventional wisdom associates fire retardancy in polymers with the presence of one or more of the following elements: Br, P, Cl, Sb, N and S (*1*). However, recently it was discovered that unsized polyester fabric can be flame retarded with certain organic compounds that are based on carbon, hydrogen and oxygen (C, H, O) only (*2,3*).

240

Experimental

Materials and Sample preparation

Unsized polyester fabric (150 g/m^2), woven from fiber supplied by South African Nylon Spinners, was used. Sample strips (50 x 200 mm) were cut and dried at ca. 70°C for a few days. The fabric strips were treated with solutions or suspensions of prospective flame retardant additives and allowed to dry at ambient conditions for 24 hours. This was followed by another oven treatment at 60°C for one day. The additive add-on level was determined by weighing the fabric after each oven drying. The add-on is defined as the mass percentage additive loading with respect to the untreated, dry fabric.

Characterization

Samples of the untreated fabric and fabric treated with pentaerythritol, dipentaerythritol, phloroglucinol and isophthalic acid were subjected to cone-and-plate viscometry, Differential Scanning Calorimetry (DSC) and Thermal Gravimetric Analysis (TGA). In each case the temperature was scanned at a rate of 5°C/min in an air atmosphere.

Flame resistance tests

The self-extinguishing times, of vertically suspended fabric samples, were determined using the NCB 245 spirit burner ignition test (4). Specimens were suspended from clamps so that they hung vertically. They were exposed, from below, to a spirit burner flame for a duration of 10 seconds. If the sample was not consumed completely, it was exposed to a second flame application of the same duration. The average self-extinguishing time for six ignitions was noted. When flame was applied, the fabric either ignited and burned or melted and dripped away from the flame front. It was therefore necessary to move the burner upward in order to maintain contact between the flame and the edge of the fabric as required by NCB 245. The mass loss and the amount of polymer that dripped to the floor were also noted.

Samples were also subjected to a face ignition test. In this test a piece of fabric was mounted on a vertical wire frame. Note that the fabric was not pulled during mounting, but only loosely fixed to the frame to prevent its natural curling effect. A horizontal gas burner flame was applied for ten seconds to the

242

middle of the fabric face and the response noted. The area of the burn-hole was also determined.

Rate of dripping

The rate of dripping was measured in real time using an Ohaus Explorer balance. The mass-time data were captured on a personal computer using appropriate software. A 50mm wide gas nozzle was used to generate a flat horizontal flame front. It had the same width as the fabric samples for good overall heating and melting. The gas flow was set at a constant rate using a bubble flow meter to ensure an even, constant heat flux.

The fabric samples were attached to a wire frame to keep them spread open. They were lowered into the fixed horizontal gas-burner flame at a rate of approximately 1.8 ± 0.2 mm/s using a motorized set-up. This speed was selected to match the burning and melting rate of the fabric.

The molten polymer that dripped down was collected on an asbestos-stainless steel plate assembly placed on top of the balance load-plate. The dripping occurred erratically. The rate of dripping was therefore estimated by fitting a straight line, passing through the origin, through the mass-dripped versus time data as illustrated in Figure 4. The rate of dripping was determined from the averaged slopes obtained from six such measurements per additive treatment.

Results and Discussion

Bottom edge ignition test

In the bottom-edge ignition test the untreated fabric usually burned out completely or burned for longer than 30 seconds. Use of volatile additives, e.g. maleic anhydride, further increased the flammability of the fabric. Table 1 lists results obtained for a series of low volatility additives. The measured self-extinguishing (SE) times for treated samples showed large scatter with standard deviations up to double the average SE time. Despite this, excellent self-extinguishing times were achieved with relatively low add-on levels for some additives. For example, dipentaerythritol required an add-on of just above two percent to be effective as a flame retardant with an average self-extinguishing time of ca. 4 seconds.

Figure 1 shows the variation of self-extinguishing time with additive loading for 2-furoic acid, benzophenone and dipentaerythritol. Interestingly, the self-extinguishing curves are parabolic. This implies that the present additives are effective in an intermediate concentration range only. Such behavior was previously also observed with a hindered N-alkoxy amine (NOR-HALS) as flame retardant in polypropylene fabric (5). Using the NFPA 701 vertical burn

Table 1: Self-extinguishing Times for the Bottom-edge Ignition Test

Functional Type	Compound	Add-on [%]	SE Time, [s] Avg.	Std. Dev.
Alcohols	pentaerythritol	7.3	4.7	4.3
	dipentaerythritol	2.1	3.8	2.7
	m-inositol	1.8	3.3	2.6
Ketones	benzyl phenyl ketone	1.8	0.5	0.7
	benzophenone	3.1	2.8	2.1
Peroxide	benzoyl peroxide	≈ 1	7	11
Anhydrides	maleic anhydride	3	>20	burns
	phthalic anhydride	1.2	1.8	2.4
Carboxylic Acids	Benzoic acid	4.6	3.3	3.1
	2-furoic acid	6.7	0.7	1.2
	terephthalic acid	1.2	0.3	0.6
	isophthalic acid	1	0	0
Epoxides	Epikote 1001	5.1	1.3	1.3
	Epikote 3004	17	0.5	0.7
	Epikote 3009	16	2.7	1.2
Esters	benzoyl benzoate	14	2.3	0.6
	diethyl phthalate	12	2.7	0.6
Phenols	pyrogallol	3	1	1
	resorcinol	5.2	7	4
	phloroglucinol	4.8	0.2	0.4
Multi-functional	maltol	3.9	3.0	4.2
	vanillin	4.8	1.8	1.0
	4-hydroxybenzoic acid	25	2.3	1.3

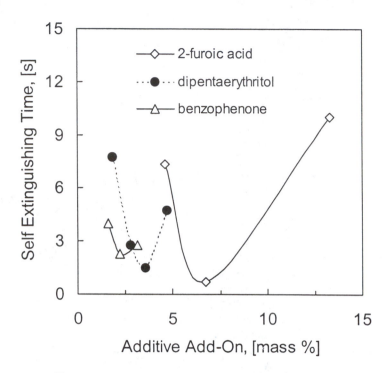

Figure 1: Self-extinguishing times for polyester fabric as a function of add-on for selected additives.

test, polypropylene fiber and films passed the test with the addition of 1% NOR-HALS. However, when the concentration of the NOR-HALS was raised to 10% the test was failed.

In some cases (e.g. pentaerythritol and phthalic anhydride) the incorporation of the additive led to profuse dripping. However, other additives (e.g. benzyl phenyl ketone, oxalic acid) did not show increased dripping.

Face Ignition Test

The following was observed during the face ignition tests: The fabric first appeared to contract at the point where the flame contacted the fabric. As melting started a hole formed rapidly. It appeared to enlarge as if pulled outward by an external force. The molten polymer formed a ridge along the edge of the hole as it expanded. It also formed large, tear shaped beads as shown in Figure 2. The untreated fabric ignited in this test. This resulted in a black char residue that was deposited on the colder surfaces by the burning process. In contrast, the treated fabric samples failed to ignite. In addition, the hole that formed in the fabric was significantly smaller than the one that burned into the untreated fabric.

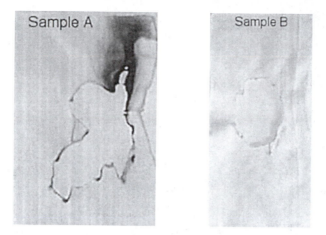

Figure 2: Face ignition test for Sample A (untreated polyester fabric) and Sample B, a fabric treated with pentaerythritol at an add-on level of 1.2%. The treated sample failed to ignite in the test and also developed a smaller hole measuring approximately 15 mm x 20 mm.

Thermal Analysis

The thermogravimetric (TG) curves for treated and untreated fabric samples were nearly identical. Mass loss started above 400°C and occurred in two stages. Approximately 80% of the mass loss took place in the first step. The second step commenced around 450°C and led to a total sample mass loss that exceeded 90% at 600°C.

In contrast, differences were observed in the DSC curves. The melting endotherm of the untreated fabric extended from 248°C to 260°C with a peak at 254°C. The presence of the additives led to a broadening of the melting range (from 247°C to as high as 268°C) while the peak temperature also shifted to higher values (up to 258°C). However, within the experimental variability of the measurements, the additives did not affect the latent heat of fusion of 52.3 ± 1.3 J/g.

Viscometry

Cone-and-plate rheometry produced a surprising result. According to the data of Figure 3, dipentaerythritol increased melt viscosity rather than inducing the expected reduction as observed for pentaerythritol.

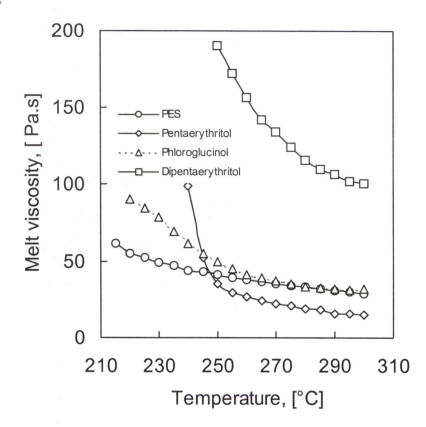

Figure 3: Effect of additives on the viscosity of the polymer melt.

Dripping Test

Figure 4 shows the results obtained from a typical drip test. The average values obtained for the polymer drip test are given in Table 2. The large scatter in the data makes comparisons difficult. However, isophthalic acid appears not to affect the dripping rate whereas pentaerythritol increases it. Fabric samples treated with the latter additive dripped almost 70% faster than the observed rate for the untreated sample.

Thermoplastics have a strong tendency to melt and drip when exposed to open flames. Flaming drips pose the hazard of spreading flames beyond the

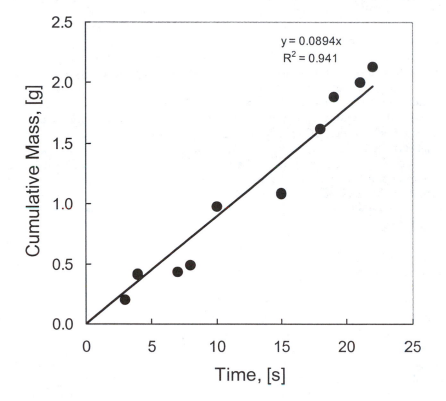

Figure 4. Typical mass loss versus time curve measured in a dripping test for a fabric treated with pentaerythritol

Table 2: Results Obtained from the Dripping Tests

Additive	Rate of Dripping, [g/s]	
	Average	*95% Confidence Interval*
Untreated	0.077	0.063 – 0.090
Pentaerythritol	0.129	0.107 – 0.151
Dipentaerythritol	0.085	0.060 – 0.111
Phloroglucinol	0.081	0.072 – 0.091
Isophthalic acid	0.070	0.051 – 0.089

initial site of ignition (*6,8*). However, some of the standard tests for evaluating fire retardancy allow dripping to occur. For example, the Underwriters Laboratory UL 94 V-2 rating makes provision for flaming drips. Gouinlock *et al.* (*7,8*) have examined the effect of dripping on the self-extinguishing properties of polystyrene and polypropylene. They used peroxides in combination with brominated flame retardants to enhance the tendency of these polymers to drip. Phosphorus compounds are also used to catalyze polyester thermal decomposition in order to induce dripping (*9*). This dripping of hot or flaming polymer removes heat from the flame zone and can contribute significantly to extinguishing the flame. The observed increase in the tendency to drip suggests that this mechanism did play a role, at least in the case of pentaerythritol.

The rate of dripping of molten polymer off a semi-solid residue is determined by a combination of the melt viscosity and the surface tension. The fact that the presence of some additives, e.g. dipentaerythritol, appears to increase melt viscosity suggests that reduction of surface tension may have played a role.

Fabric pullback

The polymer chains in the fibers are highly oriented in the fiber direction. This low-entropy state was locked-in when the chains crystallized in an extended chain configuration during the drawing step of fiber spinning. When raising the temperature to above the melting point, the chains will have a natural tendency to assume their preferred random coil conformation. This will set up an internal stress in the fibers that will cause shrinkage of the fabric away from the ignition source as observed in the face ignition test. The observed burning of the untreated fabric implies that, in this case, the fiber pullback was not fast enough to prevent ignition. It is suspected that the broadening of the melting range, in the presence of additives (e.g. pentaerythritol and phloroglucinol), may assist faster shrinkage to maintain the temperature below the ignition temperature in the case of treated samples.

A simple model

A simplified ignition temperature model postulates the existence of a critical ignition temperature, T_{crit} (*10*). The flame will self-extinguish or fail to ignite at temperatures below this threshold. For a given applied flame the heat input is constant. We assume that the heat flux supplied by the flame is sufficient to heat the untreated fabric to a temperature above T_{crit} and thus to cause ignition. One must now consider how the incorporation of the additive can lead to an increased heat loss in order to maintain the melt temperature below T_{crit}. Heat removal by dripping and by the fabric shrinking away from the applied flame are plausible explanations. The melting of the crystalline domains in the polymer also provides a heat sink effect due to the large latent heat of fusion. It is postulated that the additives function by a combination of these effects in a strongly coupled way.

Conclusions

Organic oxygen containing additives, added to unsized polyester at intermediate levels, can render the fabric "flame retarded" in terms of vertical flame tests. It is postulated that the mechanism involves both enhanced dripping and fiber shrinkage. These two effects allow efficient heat removal from the burning sample causing it to self-extinguish. Melt viscosity measurements suggest that the observed increased rate of dripping can not be ascribed to a viscosity reduction alone. It is speculated that surface tension effects might be responsible instead.

Acknowledgements

The authors express their gratitude to the Center for Polymer Technology, a joint venture between the CSIR and the Technikon of Pretoria, for access to TGA, DSC and rheology instruments. Financial support from the THRIP programme of the Department of Trade and Industry and the National Research Foundation as well as Xyris Technology is gratefully acknowledged.

References

1. Lyons, J. W. *The Chemistry and Uses of Flame Retardants*; Wiley: New York, 1970, pp. 14-17.
2. Bisschoff, J.; Focke, W.W. *Oxygenated Hydrocarbon Flame Retardants,* S.A. Patent Application No. 99/2998, **1999**.
3. Bisschoff, J. M. Eng. Thesis, *Oxygenated Hydrocarbon Flame Retardant for Polyester Fabric,* University of Pretoria, **2000**.
4. National Coal Board Specification 245: *Fire and Electrical Resistance Properties of Supported and Unsupported Sheeting,* **1985**.
5. Srinivasan, R.; Gupta, A.; Horsey, D. *A Revolutionary UV Stable Flame Retardant System for Polyolefins, Proceedings of the 1998 International*

250

Conference on Additives for Polyolefins, Soc. Plast. Eng. **1998**, Ciba Speciality Chemicals Corporation, Tarrytown, New York.

6. J Eichhorn, J. *J. Applied Pol. Sci.* **1964**, *8*, 2497-2524.
7. Gouinlock, E.V.; Porter, J.F.; Hindersinn, R.R. *Organic Coatings & Plastics Chem.* **1965**, *28*, 225-236.
8. Gouinlock E.V., Porter J.F.; Hinderson R.R. *J. Fire and Flammability*, **1971**, *2*, 206-218
9. Weil, E.D. *Mechanisms of Phosphorus-Based Flame Retardants*; FRCA Conference, Orlando, Florida, **1992**, p 19-32.
10. Kishore, K. *Pol. Eng. Rev.* **1983**, *2*, 257-293.

Halogen

The use of halogen compounds as additives or in backbone approaches to polymer flammability is an established technology. As noted in the Najafi-Mohajeri chapter on polyurethanes, halogen flame retardants may be the most effective flame retardants for some polymer systems. Flame retardants constitute 27% of the market for additives in plastics. In 1997 almost half of the dollar value of flame retardants worldwide were in diverse halogen materials, 1/5 in chlorine and 4/5 in bromine based materials. The total value of halogen flame retardant compounds was $1.1 billion (1).

Halogen-containing flame retardants are frequently used with antimony oxide (Sb_2O_3). It is well known that antimony trihalide is formed, which is volatile and plays a role, if not a predominant role, in the flame retardancy. The first paper (Starnes et al.) in this section presents a detailed mechanistic discussion of the decomposition of a specific, commonly used, chlorinated flame retardant additive in the presence of antimony oxide in nylon 6,6. The paper provides an important discussion on the uncertain area of exact mechanism.

The use of halogen flame retardants generally results in a marked increase in smoke on burning of a material containing the flame retardant. Considerable work has been done to reduce that clearly negative impact. The second paper (Pike et al.) reports on compounds that may increase crosslinking of PVC and thus reduce smoke formation. PVC is the principal plastic used in construction applications, an area where fire performance is clearly important.

The third paper (Moller et al.) addresses a topic of some controversy. In Europe particularly there is the notion that halogen flame retardants may render plastics less "environmentally friendly" and have a negative impact on recycling (2). The investigation in this paper examines high impact polystyrene (HIPS) with and without decabromodiphenyloxide as flame retardant. The results on commercial materials indicate that there is no loss of flame retardant on aging at 80°C for 1400 hours (simulated 10-year service). No degradation of flame retardant was seen. Fire performance was maintained by FR-HIPS after aging. And FR-HIPS maintained physical properties to a higher degree than HIPS itself, perhaps due to a better stabilizer package. Producers of a number of plastics have shown that flame retardant resins can go through multiple molding cycles without serious property degradation, including flammability performance (3-5). Halogen flame retarded resins are clearly recyclable, i.e., they can be collected, reground, and be remolded into functional parts.

References

1. Georlette, P.; Simons, J.; Costa, L. Halogen-Containing Fire-Retardant Compounds. *Fire Retardancy of Polymeric Materials;* Grand, A. F.; Wilkie, C. A., Eds.; Marcel Dekker: NY, 2000, pp 245-284.
2. Nelson, G. L. The Changing Nature of Fire Retardancy in Polymers. *Fire Retardancy of Polymeric Materials;* Grand, A. F.; Wilkie, C. A., Eds.; Marcel Dekker: NY, 2000, pp 1-26.
3. van Riel, H. C. H. A. In: 1994 Fall Conference Proceedings; Fire Retardant Chemical Association: Lancaster, PA, 1994, pp 167-173.
4. Bopp, R. C. In: 1994 Fall Conference Proceedings; Fire Retardant Chemical Association: Lancaster, PA, 1994, addendum.
5. Christy, R.; Gavik, R. In: 1994 Fall Conference Proceedings; Fire Retardant Chemical Association: Lancaster, PA, 1994, pp. 151-166.

Chapter 20

Reductive Dechlorination of a Cycloaliphatic Fire Retardant by Antimony Trioxide and Nylon 6,6: Implications for the Synergism of Antimony and Chlorine

William H. Starnes, Jr., Yun M. Kang, and Lynda B. Payne

Departments of Chemistry and Applied Science, College of William and Mary, P.O. Box 8795, Williamsburg, VA 23187-8795

In nylon 6,6 containing Sb_2O_3, the Diels-Alder adduct made from hexachlorocyclopentadiene (two equivalents) and 1,5-cyclooctadiene experiences partial reductive dechlorination at 320–330 °C. This reaction is accompanied by a weight loss that apparently results, in part, from the volatilization of HCl and/or the gas-phase fire retardant, $SbCl_3$. Thus, at higher temperatures, the reaction seems likely to play a major role in the antimony/chlorine synergism that suppresses flame in this system. Possible mechanisms for the reductive dechlorination are described.

Mixtures of chloro- or bromoorganics with antimony trioxide (Sb_2O_3, commonly called "antimony oxide") have been widely used for many years as fire retardants for polymers. The predominant feature of their performance is the synergism that arises from their generation *in situ* of volatile antimony halides that act as flame suppressants in the vapor phase (*1–3*). Surprisingly,

254

the reactions that form those substances still are not well-understood. Condensed-phase interactions of antimony/halogen systems might also inhibit flame by reducing the formation of volatile fuel. However, this type of inhibition frequently seems to be much less important, as in the case of (antimony oxide)/poly(vinyl chloride) (PVC) blends (*4–6*).

This chapter reports a preliminary investigation of the condensed-phase chemistry that is operative in heated mixtures of antimony oxide, nylon 6,6, and compound **1**, a well-known fire retardant for polymers (*7*). This work has revealed the occurrence of an unexpected reaction that seems likely to play a major role in the generation of SbCl₃.

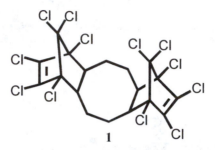

1

Background

Vapor-Phase Fire Retardance by Antimony Trichloride

In antimony/chlorine-containing polymer systems, SbCl₃ ordinarily is regarded as the actual fire retardant (*2,3,8*). Yet in one recent study, Shah et al. (*9*) observed no SbCl₃ in the volatile pyrolysate formed from antimony oxide and PVC at 1000 °C. They detected SbCl₂, instead, but pointed out that their experimental methodology (flash pyrolysis and analysis by chemical ionization mass spectrometry) might not have been representative of actual fire situations (*9*).

Costa and co-workers (*10*) have provided evidence for the operation, in polypropylene, of a synergistic retardance process in which certain metal chlorides inhibit fuel volatilization by oxidizing macroradicals. Reoxidation of the resultant reduced-metal species by an admixed chloroparaffin then re-creates the metal chloride (*10*). However, this condensed-phase redox mechanism apparently is inoperative with Sb₂O₃ (*8,10*), a result that can be explained, at least in part, by the very rapid volatilization of antimony trichloride at the temperatures of burning polymers [typically, 400–700 °C (8)].

Possible Routes from Antimony Oxide to Antimony Trichloride

The mechanism for the formation of $SbCl_3$ is controversial. Several types of mechanisms have, in fact, been proposed. One of them pertains to the use of chloroorganics that can easily liberate HCl at combustion temperatures. The HCl may react with antimony oxide to form $SbCl_3$, either without the intervention of observable intermediates (eq 1) (11) or via an intermediate oxychloride such as SbOCl (eqs 2 and 3) (11).

$$Sb_2O_3 + 6HCl\!\uparrow \longrightarrow 2SbCl_3\!\uparrow + 3H_2O\!\uparrow \qquad (1)$$

$$Sb_2O_3 + 2HCl\!\uparrow \longrightarrow 2SbOCl + H_2O\!\uparrow \qquad (2)$$

$$SbOCl + 2HCl\!\uparrow \longrightarrow SbCl_3\!\uparrow + H_2O\!\uparrow \qquad (3)$$

Another type of mechanism relates to chloroorganics such as 1 that cannot undergo the facile thermal loss of HCl. With these substances, the route to $SbCl_3$ obviously must begin with the direct transfer of chlorine to antimony, and numerous ways of accomplishing such a transfer have been suggested. They include (a) the direct attack of a C–Cl moiety by Sb_2O_3 (12), (b) the reaction of Sb_2O_3 with chlorine-containing species (other than HCl) that result from the thermolysis of the organic chloride (12), (c) oxidation of the polymer by Sb_2O_3 to form Sb(0), which then acquires chlorine from a C–Cl bond (13) [cf. (10)], and (d) the abstraction of chlorine from such a bond by an (antimony oxide)/polymer complex (13). Subsequent transfers of chlorine can be envisaged to occur in similar ways. However, an alternative possibility (which also applies to systems where HCl is formed) is that $SbCl_3$ results from the thermal disproportionation of oxychloride intermediates. A disproportionation pathway that currently seems acceptable appears in eqs 4–6 (8).

$$5SbOCl \xrightarrow{270\text{-}275\ °C} Sb_4O_5Cl_2 + SbCl_3\!\uparrow \qquad (4)$$

$$11Sb_4O_5Cl_2 \xrightarrow{405\text{-}475\ °C} 5Sb_8O_{11}Cl_2 + 4SbCl_3\!\uparrow \qquad (5)$$

$$3Sb_8O_{11}Cl_2 \xrightarrow{475\text{-}570\ °C} 11Sb_2O_3 + 2SbCl_3\!\uparrow \qquad (6)$$

A third type of mechanism for antimony trihalide formation also involves certain additives that do not give hydrogen halide directly (*13*). They are proposed to undergo reactions with macroradicals to produce free halogen atoms that then abstract hydrogen from the polymer in order to form the hydrogen halide that reacts with antimony oxide (*13*).

The literature that relates to the formation of $SbCl_3$ is extensive. Thus we will make no attempt to review it exhaustively here. Instead, we will simply call the reader's attention to some published observations that seem to relate rather closely to our present work. One of these is the demonstration by Gale (*14*) that antimony trihalide formation from Sb_2O_3 benefits greatly from the presence of labile hydrogen as well as organically bonded halogen.

Another pertinent inquiry (*15*) used thermogravimetric analysis (TGA) to compare the pyrolysis characteristics of mixtures of antimony oxide with two organic chlorides at 200–800 °C. When HCl could be formed directly from the chloride, the Sb_2O_3 was converted into $SbCl_3$ without the noticeable incursion of reactions of intermediate oxychlorides (*15*). On the other hand, when the dehydrochlorination of the starting chloride was structurally unattainable, the weight loss profile was consistent with the initial formation of SbOCl and its later thermal conversion into $SbCl_3$ and antimony oxide via a series of steps (*15*) (cf. reactions 4–6).

In an especially relevant study, both $SbCl_3$ and other chlorine-containing (presumably organic?) species were found to be evolved when Sb_2O_3 was heated with compound **1** at ca. 400–700 °C (*8*). However, the researchers (*8*) reported no experiments with ternary mixtures containing **1**, Sb_2O_3, and a combustible polymer.

Research Objectives

The present work was concerned with the pyrolysis chemistry of mixtures of nylon 6,6 with compound **1** and antimony oxide. Weight losses from such mixtures were determined by TGA, and the pyrolysis products were identified primarily by (gas chromatography)/(mass spectrometry) (GC/MS). Quantitative product analysis was not attempted, because some of the products could not be characterized, and the GC sensitivity factors of several others could not be measured, owing to the unavailability of the pure materials.

One objective of the research was to understand the initial interaction of antimony oxide with the chlorinated fire retardant. For that reason, rather low pyrolysis temperatures were used in an attempt to minimize complications.

Experimental

Materials

Antimony oxide, compound **1**, nylon 6,6, and polymer/additive mixtures prepared by molding were supplied by the Occidental Chemical Corporation (OxyChem). The other chemicals were obtained from various commercial suppliers. They had the highest available purities and were used as received.

Instrumental Analysis

Thermogravimetric analysis (TGA) was carried out with a Seiko SSC 5040 system incorporating a TG/DTA 200 analyzer with Version 2.0 software. The samples used (12–20 mg) were taken from the molded blends supplied by OxyChem or from mixtures prepared by dry blending with a mortar and pestle at liquid nitrogen temperature. Heating was conducted under nitrogen (50 mL/min) at the rate of 10 °C/min up to 320 °C, and that temperature then was maintained for 30 min prior to cooling.

(Gas chromatography)/(mass spectrometry) (GC/MS) analyses were performed with a Hewlett-Packard apparatus (Model 5890/5971A) equipped with a fused-silica HP-1 capillary column [12 m × 0.2 mm (i.d.)] and Hewlett-Packard G1034B software for the MS ChemStation (DOS Series). The carrier gas was helium, and the temperature of the injection port was 200 °C. Column temperature was increased from 50 to 300 °C at a programmed rate and then was held at 300 °C for 10 min. Where possible, products were identified by comparing their GC retention times and mass spectra with those of authentic specimens. Identification of the chlorinated products was aided considerably by comparisons of the relative abundances of their mass peaks with those expected for various ions whose chlorine isotopes differed (*16*).

A General Electric QE-300 NMR spectrometer and a Perkin-Elmer 1600 Series FTIR instrument also were used for product characterizations. The FTIR samples were examined in KBr pellets.

Pyrolysis Product Identification

Pyrolysis of the additives, the polymer, and their various mixtures were performed on a 2.0-g (total weight) scale for 30 min at 325±5 °C under flowing nitrogen. Volatile products were trapped at –85 °C and dissolved in

258

tetrahydrofuran (THF) before GC/MS analysis. Other GC/MS analyses were performed on THF extracts of the pyrolysis residues that had not volatilized. Compounds (e. g., **1**) derived from one molecule of 1,5-cyclooctadiene (**2**) and two molecules of chlorinated cyclopentadienes usually did not exhibit parent-ion mass spectral peaks. These compounds were identified by their lengthy GC retention times and by their apparent partial conversion, in the mass spectrometer, into the cyclopentadienes that would have resulted from retro-Diels-Alder reactions.

Reactions of Hexachlorocyclopentadiene (3) with Antimony Oxide

These reactions were carried out in glass tubes that had been sealed under argon with rigorous exclusion of air and mosture. The tubes were heated at 250±10 °C for various lengths of time.

Preparation of the Diels-Alder Monoadduct, 4

In an adaptation of a published procedure (*17*), 2.68 g (24.8 mmol) of diene **2** and 1.70 g (6.23 mmol) of cyclopentadiene **3** were heated together under nitrogen for 4 h at 100±1 °C. Analysis of the resultant mixture by GC/MS revealed the presence of substantial amounts of unchanged **2** and **3**, together with a major GC peak that was identified from its mass spectrum as monoadduct **4**. In addition to weak parent ions at m/e 378, 380, 382, and 384 ($C_{13}H_{12}Cl_6$), the spectrum showed stronger fragments at m/e 343, 345, 347, and 349 ($C_{13}H_{12}Cl_5$).

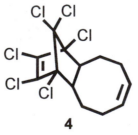

4

Results

Table I shows total weight losses, as determined by TGA, for nylon 6,6, the two additives, and several mixtures thereof. The tabulated values are the averages of those from a number of replicate runs, and the data for mixtures refer to dry blends prepared by us, for which the deviations generally amounted to ±5–15% of the totals listed. The molded blends supplied by OxyChem showed equally good reproducibility but tended to give loss values that were higher by a few percent.

Table I. TGA Weight Loss Study[a]

	wt %			wt loss, %		
Run	Nylon 6,6	1	Sb$_2$O$_3$	Theoret	Found	Dev, %
1	100				6	
2		100			52	
3			100		0	
4	90	10		11	15	36
5	90		10	5	5	0
6		90	10	47	41	(13)[b]
7	88	10	2	11	16	45
8	84	10	6	10	19	90
9	76	19	5	14	26	86
10	72	18	10	14	28	100

[a]10 °C/min to 320 °C, 30 min hold. [b]Negative value.

Pure nylon 6,6 (run 1) produced only a small weight loss that was shown to consist, at least partially, of cyclopentanone, as expected (18). No attempts were made to identify the anticipated gaseous products (NH$_3$, H$_2$O, and CO$_2$) (18) having lower molecular weights. In contrast, compound **1** (run 2) lost a considerable amount of weight, a result that was ascribable primarily to sublimation, in that the volatile product was a white solid whose decomposition point (ca. 350 °C) and FTIR spectrum were essentially the same as those of the starting **1**. Run 3 confirmed the expected nonvolatilization of antimony oxide at the temperatures used.

For runs 4–10, the theoretical weight losses in the table are the values predicted from runs 1–3 by arithmetic additivity. The deviations in the last column represent the differences between the theoretical and actual losses (Found − Theoret), and they are reported as percentages of the theoretical values.

Run 4 shows a mass loss increase which, though rather small, seems statistically valid. In the corresponding preparative pyrolysis, all of the volatiles could not be identified, but two of the major ones were shown to be diene **2** and the monoadduct (**4**) that would have resulted from the retro-Diels-Alder sequence shown in eq 7. The GC retention time and mass spectrum of **4** were identical to those of an authentic specimen prepared directly from **2** and **3**. Under the conditions of run 4, the retro-Diels-Alder reactions should have been favored by the dilution effect of the polymer, which would have retarded the back reactions in eq 7 by reducing the concentrations of compounds **2–4**. Interestingly, **3** was not actually detected as a pyrolysis product, presumably because it was diverted into insoluble and nonvolatile (polymeric?) material.

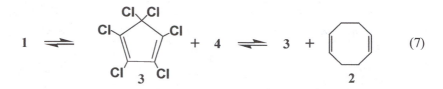

The most interesting feature, however, of the preparative pyrolysis analogous to run 4 was the formation of two volatile products (probably stereoisomers) whose elemental composition was $C_{13}H_{13}Cl_5$. This result and their mass spectral cracking patterns indicated that these compounds were analogues of **4** in which one of the chlorines had been replaced by hydrogen.

For purposes of comparison, pure **1** was heated in a sealed tube under argon for 22 h at 320–370 °C. Under those conditions, the isomer composition of the substance changed dramatically (from a ratio of ~2:5 to a ratio of ~4:1), and several lighter products were formed. They included **3** (in low yield), **4**, pentachlorobenzene, tetrachloroethylene, and two polychlorides whose lowest parent-mass values were 364 and 376. These results cannot be interpreted fully at this time. Nevertheless, they reaffirm the occurrence of retro-Diels-Alder reactions in this system and provide further indirect evidence for the thermal conversion of compound **3** into insoluble product(s), in that the yield of **3** was much less than that of **4**. The change observed in the isomer ratio of **1** could signify an approach to the equilibrium composition via reaction sequence 7. On the other hand, the change in ratio might have resulted, instead, from the selective and irreversible conversion of one isomer into non-Diels-Alder products.

The most striking results in Table I are those for runs 7–10. These data show that ternary mixtures of the polymer, **1**, and Sb_2O_3 underwent total weight losses that were much greater than those expected on the basis of additivity. The observed weight losses were larger when the Sb_2O_3:**1** weight ratio was increased. Moreover, plots (not shown) of weight loss vs time showed that antimony oxide also increased the initial weight-loss rate. In analogous preparative runs, cyclopentanone and diene **2** were identified as products. The major products, however, were substances that would have resulted from the partial reductive dechlorination of compounds **1** and **4**. Those substances are identified in Figures 1 and 2, which show portions of the gas chromatograms of the product fractions obtained from a ternary mixture similar to that in run 8. Remarkably, no unreduced **4** ($C_{13}H_{12}Cl_6$) appears in either chromatogram, and the amount of unchanged **1** (both isomers) is much less than the total yield of dechlorinated analogues, which have lost up to four chlorines. Thus it is clear that the reductive dechlorination is a facile process, even at the relatively low temperature of ca. 325 °C.

Discussion

When antimony oxide is not present (as in run 4), the occurrence of reductive dechlorination can be explained by C-Cl homolysis, followed by hydrogen abstraction from the polymer by the carbon-centered radicals that are formed. Indeed, no reasonable alternative rationalization is apparent at this time. The other homolysis product, Cl·, is an extremely reactive radical that also is likely to abstract hydrogen, perhaps to a major extent. Thus the formation of the gaseous substance, HCl, can account for some of the increased weight loss.

The much larger losses in weight caused by antimony oxide could be due to the increased evolution of HCl and organic products, and/or to the evolution of $SbCl_3$. Low solubility of the latter substance in the extraction solvent (THF) would have tended to prevent its detection. However, when mixtures of Sb_2O_3 and the retro-Diels-Alder product, **3**, were heated at 250±10 °C under argon for various lengths of time, $SbCl_3$ was formed as a crystalline sublimate and identified conclusively by its mass spectrum. Thus, in runs 7–10, $SbCl_3$ formation does not seem unlikely. It could have resulted from various combinations of reactions 1–4.

At present, we do not know whether the reaction of Sb_2O_3 with **1** and nylon 6,6 involves free radicals or ionic intermediates. Nevertheless, as noted above, a free-radical mechanism seems very probable for the small amount of reductive dechlorination that occurred under the conditions of run 4. Equations 8–11 represent a simple but speculative free-radical scheme that accounts for enhanced reductive dechlorination and greater weight loss when antimony oxide is present. According to this mechanism, the Sb_2O_3 acts as a catalyst for both processes. Some of the Cl· radicals may attack the nylon, of course, but reaction 9 does not need to be quantitative in order to account for catalysis. The HCl that results from reaction 11 may either volatilize as such or react with antimony oxide to initiate the formation of volatile $SbCl_3$. Partial reductive

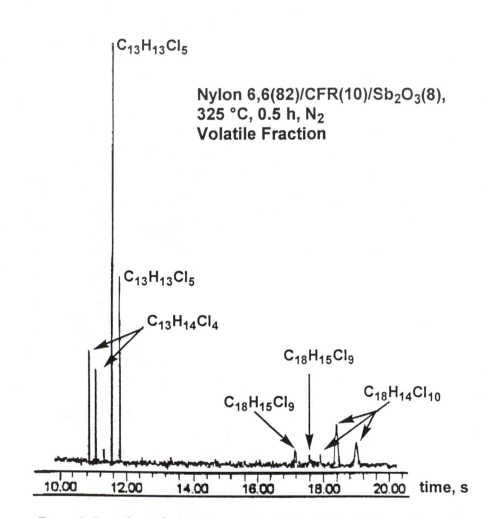

Figure 1. Partial gas chromatogram of the THF-soluble volatile products of a pyrolysis described in the heading. Parenthesized numbers in the heading are percentages by weight; chlorinated fire retardant 1 is designated as "CFR".

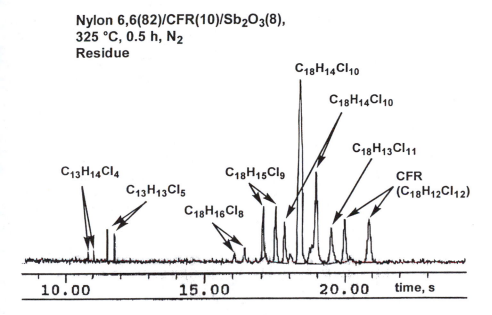

Figure 2. Partial gas chromatogram of the THF-soluble residue from a pyrolysis described in the heading. Parenthesized numbers in the heading are percentages by weight; chlorinated fire retardant 1 is designated as "CFR".

dechlorination will significantly enhance the vaporizabilities of several organic products. Alternatively, a direct attack of a C-Cl group by antimony oxide could also lead to catalysis, and the continual repetition of the entire scheme will obviously give the products whose dechlorination is more extensive.

$$1 \quad \xrightarrow{\Delta} \quad C_{18}H_{12}Cl_{11}^{\bullet} \;+\; Cl^{\bullet} \tag{8}$$

$$Cl^{\bullet} \;+\; Sb_2O_3 \;\rightleftharpoons\; ClSb_2O_3^{\bullet} \tag{9}$$

$$C_{18}H_{12}Cl_{11}^{\bullet} \;+\; -NHCH_2CH_2- \;\longrightarrow\; C_{18}H_{13}Cl_{11} \;+\; -NH\overset{\bullet}{C}HCH_2- \tag{10}$$

$$ClSb_2O_3^{\bullet} \;+\; -NH\overset{\bullet}{C}HCH_2- \;\longrightarrow\; -NHCH=CH- \;+\; HCl \;+\; Sb_2O_3 \tag{11}$$

Ionic mechanisms for the dechlorination can be written, as well. Moreover, consideration can be given to the possibility that the initial attack by antimony oxide occurs on compound 3 or on chlorine atoms that result from its homolysis. An especially intriguing speculation is that antimony oxide may be sufficiently nucleophilic to abstract a chloronium cation from a bridge carbon of 1 or 4, as in eq 12. In that event, the retro-Diels-Alder reactions of the resultant anions (e. g., eq 13) would be exceptionally fast, because they would form an aromatic species, $C_5Cl_5^-$. Numerous precedents for the great rapidity of similar reactions are available (*19–22*).

$$Sb_2O_3 \;+\; \underset{\mathbf{1}}{(C_{17}H_{12}Cl_{10})CCl_2} \;\rightarrow\; ClSb_2O_3^+ \;+\; (C_{17}H_{12}Cl_{10})CCl^- \tag{12}$$

$$(C_{17}H_{12}Cl_{10})CCl^- \;\rightarrow\; \mathbf{4} \;+\; C_5Cl_5^- \tag{13}$$

Summary and Conclusions

The exploratory study described in this chapter has shown that nylon 6,6 can reductively dechlorinate compound 1 in a process that is strongly promoted by Sb_2O_3. This process occurs at lower temperatures than those encountered in fires, and it leads to losses of mass that are likely to consist, in part, of HCl and/or the gas-phase fire retardant, $SbCl_3$. Hydrogen chloride, antimony trichloride, and antimony oxychlorides would seem to be the only reasonable dechlorination should make a major contribution to the antimony/chlorine synergism that suppresses flame in this system. Further studies of the mechanism for the dechlorination clearly would be worthwhile.

Acknowledgment

We thank R. L. Markezich for useful discussions and R. F. Mundhenke for the preparation of molded specimens. This research was supported in part by the Occidental Chemical Corporation.

References

1. Hastie, J. W. *J. Res. Natl. Bur. Stand., Sect. A* **1973**, *77*, 733–754.
2. Cullis, C. F.; Hirschler, M. M. *The Combustion of Organic Polymers*; Oxford University Press: New York, 1981; pp 276–295.
3. Camino, G.; Costa, L.; Luda di Cortemiglia, M. P. *Polym. Degrad. Stab.* **1991**, *33*, 131–154.
4. Lum, R. M. *J. Appl. Polym. Sci.* **1979**, *23*, 1247–1263.
5. Starnes, W. H., Jr.; Edelson, D. *Macromolecules* **1979**, *12*, 797–802 (see also references therein).
6. Lum, R. M.; Seibles, L.; Edelson, D.; Starnes, W. H., Jr. *Org. Coat. Plast. Chem.* **1980**, *43*, 176–180.
7. Markezich, R. L.; Mundhenke, R. F. In *Chemistry and Technology of Polymer Additives*; Al-Malaika, S., Golovoy, A., Wilkie, C. A., Eds.; Blackwell Science: Malden, MA, 1999; pp 151–181.
8. Costa, L.; Goberti, P.; Paganetto, G.; Camino, G.; Sgarzi, P. *Polym. Degrad. Stab.* **1990**, *30*, 13–28.
9. Shah, S.; Davé, V.; Israel, S. C. *ACS Symp. Ser.* **1995**, *No. 599*, 536–549.
10. Costa, L.; Luda, M. P.; Trossarelli, L. *Polym. Degrad. Stab.* **2000**, *68*, 67–74.
11. Lum, R. M. *J. Polym. Sci., Polym. Chem. Ed.* **1977**, *15*, 489–497.
12. Brauman, S. K. *J. Fire Retard. Chem.* **1976**, *3*, 117–137.
13. Drews, M. J.; Jarvis, C. W.; Lickfield, G. C. *ACS Symp. Ser.* **1990**, *No. 425*, 109–129 (see also references therein).
14. Gale, P. J. *Int. J. Mass Spectrom. Ion Processes* **1990**, *100*, 313–322.
15. Avento, J. M.; Touval, I. *Kirk-Othmer Encycl. Chem. Technol., 3rd Ed.* **1980**, *10*, 355–372.
16. Beynon, J. H. *Mass Spectrometry and Its Applications to Organic Chemistry*; Elsevier: New York, 1960; pp 298–299.

17. Ziegler, K.; Froitzheim-Kühlhorn, H. *Justus Liebigs Ann. Chem.* **1954**, *589*, 157–162.
18. Levchik, S. V.; Weil, E. D.; Lewin, M. *Polym. Int.* **1999**, *48*, 532–557 (see also references therein).
19. Finnegan, R. A.; McNees, R. S. *J. Org. Chem.* **1964**, *29*, 3234–3241.
20. Bowman, E. S.; Hughes, G. B.; Grutzner, J. B. *J. Am. Chem. Soc.* **1976**, *98*, 8273–8274.
21. Neukam, W.; Grimme, W. *Tetrahedron Lett.* **1978**, 2201–2204.
22. Blümel, J.; Köhler, F. H. *J. Organomet. Chem.* **1988**, 340, 303–315.

Chapter 21

New Copper Complexes as Potential Smoke Suppressants for Poly(vinyl chloride): Further Studies on Reductive Coupling Agents

Robert D. Pike[*], William H. Starnes, Jr.[*], Jenine R. Cole, Alexander S. Doyal, Peter M. Graham, T. Jason Johnson, Edward J. Kimlin, and Elizabeth R. Levy

Department of Chemistry, College of William and Mary, P.O. Box 8795, Williamsburg, VA 23187-8795

Complexes of Cu(I) with hindered arylphosphites and salts of Cu(II) with phenylphosphonates and -phosphinates were prepared and shown to cross-link poly(vinyl chloride) (PVC) in pyrolysis studies. The use of hindered phosphites prevented the facile hydrolysis of the potential PVC additives. Complexes of Cu(I) with thioureas, nitrogen-bridged aromatics, thiophosphate, and phosphine sulfide ligands were also prepared and considered for use as additives for PVC.

Introduction and Background

Poly(vinyl chloride) (PVC) is a polymer of immense commercial importance, especially in long-term applications, e. g., use in construction. As a result, the behavior of PVC materials during fires is of importance. The thermal decomposition pathways for PVC are well-established *(1–4)*. Pyrolysis of PVC causes dehydrochlorination reactions that generate polyene segments and HCl gas, as shown in Scheme 1 *(5)*. The presence of cis linkages within the polyene segments

Scheme 1

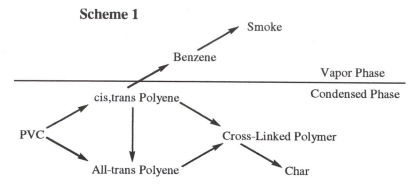

allows cyclization that produces benzene and other aromatic hydrocarbons. These hydrocarbons serve as vapor-phase fuel, thereby generating smoke. The production of cross-linked polymer is thought to be essential to char formation, which serves to keep most of the mass in the condensed phase.

Polymer additives based on high-valent metal and metalloid compounds have been widely used to help suppress smoke in PVC. Such compounds are usually strong Lewis acids, and, as a result, they catalyze Friedel-Crafts reactions, which provide the desired polyene cross-linking action. Unfortunately, under conditions of high enthalpy input, these additives can also promote the cationic cracking of the polymer char *(4–6)*. Char cracking is known to produce vapor-phase aliphatic hydrocarbons, which function as superior fuels *(5)*, burning efficiently and aiding flame spread. Hence, Lewis-acid additives that limit smoke-producing polyene cyclization reactions through cross-linking catalysis can actually increase flame in large fires.

Research in our laboratory *(4)* has demonstrated that the promotion of thermal PVC cross-linking can be accomplished through the use of low-valent metal compounds. These findings were initially spurred by the known reductive coupling of allylic halides by zero-valent metal powders *(7–9)* and low-valent metal complexes *(10–13)* according to reaction 1 (X = Cl, Br, I). (Oxidation of the metal

$$2 \; RCH{=}CH{-}CH_2X + M(0) \rightarrow RCH{=}CH{-}CH_2{-}CH_2{-}CH{=}CHR + MX_2 \quad (1)$$

serves as the thermodynamic driving force for this reaction.) It is probable that similar reductive coupling occurs during the cross-linking of PVC by low-valent metals, as indicated in Scheme 2. Although allylic chloride functionalities are

Scheme 2

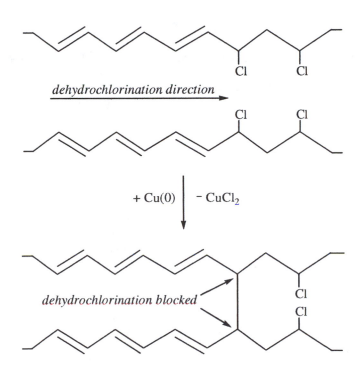

produced in PVC through thermal dehydrochlorination, a low concentration of these defect structures is actually present in the virgin polymer *(3)*. Since allylic chloride groups are found at the termini of growing polyene chains in dehydrochlorinating PVC, the dehydrochlorination process is expected to be blocked by reductive coupling at those sites. Our research has shown that freshly generated zero-valent metal films induce the coupling of allylic chloride model compounds for PVC. Similarly, slurries of freshly generated metal cause the gelation of PVC (an indication of cross-linking).

The current focus of our research has been on identifying metal compounds that will produce zero-valent metal at elevated temperatures. Such compounds would be good additive candidates, since they would produce the active cross-linking agent only under pyrolytic conditions. Herein, we report new copper complexes that are potential smoke-suppressant additives.

Copper complexes are promising for several reasons. Copper has three common oxidation states, all of which lie close together in energy. Copper(0) is relatively stable, often more so than Cu(I). This stability difference leads to disproportionation reaction 2. Copper(II) is also readily reduced. As indicated for

$$2 \, Cu(I) \rightarrow Cu(II) + Cu(0) \tag{2}$$

copper(II) oxalate in reaction 3, oxidizable ligands can drive this process. Another

$$CuC_2O_4 \rightarrow Cu(0) + 2 \, CO_2 \tag{3}$$

advantage of copper(I) among the transition metals is that its compounds are usually colorless. This property is observed because the metal has a d^{10} electronic configuration and, therefore, is not subject to electronic transitions within the d-shell. Finally, copper compounds are of minimal toxicity (copper is an essential trace metal for most organisms), and they are inexpensive.

The favorability of reactions such 2 and 3 is dependent upon the temperature and the use of supporting ligands. Thus, through judicious choice of copper complexes, we hope to develop smoke-suppressant additives that will release copper metal at temperatures well above those of polymer processing, but below those of a full-scale fire.

Experimental

General

All syntheses were carried out under a nitrogen atmosphere. All ligands, $CuSO_4 \bullet 5H_2O$, $CuCl_2 \bullet 2H_2O$, $Cu(C_2O_4)$, and $Cu(O_2CH)_2$ were used as received. The CuCl and CuBr were freshly recrystallized from aqueous HCl or HBr. The $CuCO_3$ was freshly prepared from $CuSO_4 \bullet 5H_2O$ and Na_2CO_3. Preparation of $P(OPh)_3$ (14,15) and the nitrogen-bridged (16,17) complexes of CuBr and CuCl are reported elsewhere. Details concerning PVC gelation testing, flame atomic absorption spectroscopy (AAS), and thermogravimetric (TGA) analyses are reported elsewhere (4,16).

Preparation of [CuCl(BPP)] and Other Copper(I) Complexes

Copper(I) chloride (1.00 g, 10.1 mmol) and tris(2,4-di-*t*-butylphenyl) phosphite (6.54 g, 10.1 mmol) were suspended in 50 mL of $CHCl_3$. The mixture was stirred at 25 °C under N_2 for 0.5 h, during which time it became less cloudy. After filtration, the solvent was removed under vacuum to leave a colorless oil, which solidified to a white solid under high vacuum (6.35 g, 8.51 mmol, 84%). Other complexes of CuCl and CuBr with BPP (in $CHCl_3$), MBPOP (in C_6H_6), BHTPD (in CH_3CN), CPD (in CH_3CN), Dmtu (in CH_3CN), and Etu (in H_2O) were prepared similarly (see Charts 1 and 3 and Tables I and III).

Preparation of [Cu(H$_2$PO$_2$)$_2$] and Other Copper(II) Complexes

Phosphinic acid, H_3PO_2 (1.7 mL of 50% aq. solution, 1.08 g, 16.4 mmol), was added dropwise to a suspension of $CuCO_3$ (1.00 g, 8.09 mmol) in 15 mL of H_2O at 0 °C. The resulting blue solution yielded crystals upon the addition of 100 mL of EtOH and cooling to −5 °C. The precipitate was collected by filtration, washed with acetone and ether in succession, and dried under vacuum (1.18 g, 6.07 mmol, 75%). Other Cu(II) complexes of HPO_3^{2-}, $PhPO_3^{2-}$, $HPhPO_2^-$, and $Ph_2PO_2^-$ were prepared similarly (see Table IV).

Preparation of [Cu(S$_2$P(OEt)$_2$)]

The dithiophosphate $NH_4S_2P(OEt)_2$ (2.56 g, 12.6 mmol) was dissolved in 50 mL of H_2O at 25 °C, and CuCl (1.25 g, 12.6 mmol) was added. Aqueous NH_3 was added until the green color was discharged and a white precipitate remained suspended in a blue solution. The mixture was stirred overnight, and the white product was collected via filtration. It was washed with H_2O, EtOH, and ether in succession and vacuum-dried (1.04 g, 4.18 mmol, 79%) *(18)*.

Preparation of [Cu(S$_2$P(OEt)$_2$)(SPPh$_3$)$_2$]

A suspension of $Cu(S_2P(OEt)_2)$ (0.200 g, 0.804 mmol) and Ph_3PS (0.473 g, 1.61 mmol) in 50 mL of CH_2Cl_2 was heated under reflux for 15 min. The resulting colorless solution was concentrated to a volume of 5 mL under vacuum. A white powder was precipitated by the addition of ether and vacuum-dried (0.487 g, 0.582 mmol, 72%).

Preparation of [(CuCl(SPPh₃)₃]

Addition of CuCl₂•2H₂O (0.386 g, 2.26 mmol) to a suspension of Ph₃PS (2.00 g, 6.79 mmol) in EtOH gave a green solution. A solution of SnCl₂ in EtOH was added until the green color was discharged, and a white precipitate formed on cooling overnight. It was washed in succession with EtOH and ether and vacuum-dried (1.48 g, 1.50 mmol, 66%) *(19)*.

Results and Discussion

Copper(I) Phosphite Complexes

Previous results obtained in our laboratory indicated that although copper(I) complexes of both triphenyl phosphite and triphenylphosphine are potent gel formers in pyrolyzing PVC, the phosphine complexes are unacceptable for use, owing to their promotion of dehydrochlorination *(4)*. Unfortunately, P(OPh)₃ also has the problem of being subject to facile hydrolysis. For these reasons, new Cu(I) complexes were prepared by using commercially available hindered phosphites (Chart 1) *(20)*.

Copper analysis confirmed the product stoichiometries indicated in Table I. Although the structures of the new copper(I) phosphite complexes have not yet been determined, proposed structures are shown in Chart 2. The relatively small P(OPh)₃ is known to produce "cubane" tetramers with CuX

Table I. Copper(I) Organophosphite Complexes Prepared

Complex	Color	Cu, Theory	Cu, Expt.	Dec. Point
CuCl(P(OPh)₃)	white	15.5%	15.1%	285 °C
CuBr(P(OPh)₃)	white	14.0	13.6	251
CuCl(BPP)	white	8.5	8.3	226
CuBr(BPP)	white	8.0	7.9	238
CuCl(MBPOP)	white	9.3	9.8	220
CuBr(MBPOP)	white	8.7	9.2	230
(CuCl)₂(BHTPD)	cream	15.3	15.6	235
(CuBr)₂(BHTPD)	tan	13.8	13.4	248
CuCl(CPD)	white	6.7	6.6	269
CuBr(CPD)	white	6.4	6.8	269

Chart 1

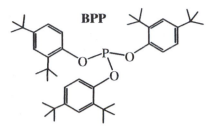

BPP

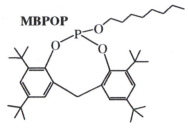

MBPOP

tris(2,4-di-*t*-butylphenyl) phosphite
(Doverphos S-480)

2,2'-methylenebis(4,6-di-*t*-butylphenyl)-
octyl phosphite   (Amfine HP-10)

BHTPD

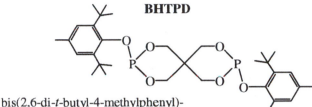

bis(2,6-di-*t*-butyl-4-methylphenyl)-
pentaerythritol diphosphite (Amfine PEP-36)

CPD

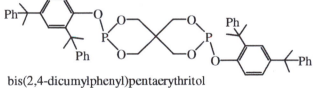

bis(2,4-dicumylphenyl)pentaerythritol
diphosphite   (Doverphos S-9228)

(15). However, the bulky monophosphites (P = BPP, MBPOP) may form dimers (lacking the "dashed" bonds, Chart 2). The two pentaerythritol diphosphite complexes (PP = BHTPD, CPD) have differing stoichiometries. Thus, [(CuX)$_2$(BHTPD)] forms polymeric chains, as shown in Chart 2, possibly with the CuX cross-linking shown. The bulkier diphosphite complexes [CuX(CPD)] are probably dimeric but may link into sheet structures, as shown.

As can readily be seen from Table II, the hindered phosphites form more hydrolytically stable complexes with copper. These data are in accord with results for the free phosphites *(20)*. Thus, coordination to copper apparently does not greatly affect the hydrolytic stability of the phosphites.

Chart 2

[CuXP]; X = Br, Cl; P = P(OPh)₃, BPP, MBPOP

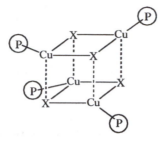

[(CuX)₂PP]; X = Br, Cl; PP = BHTPD

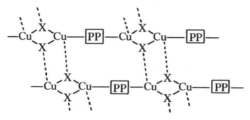

[CuX(PP)]; X = Br, Cl; PP = CPD

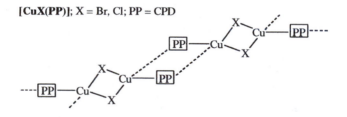

Copper(I) Sulfur and Nitrogen Complexes

Complexes of Cu(I) with the ligands shown in Chart 3 have been prepared. Copper analysis confirmed the product stoichiometries indicated in Table III. Of the numerous complexes incorporating thiourea ligands (only a few of which are listed), all were found to decompose below 200 °C, a property rendering them unsuitable for consideration as smoke-suppressant additives. The dithiophosphate complexes, Cu(S₂P(OEt)₂) and [Cu(S₂P(OEt)₂)(SPPh₃)₂], also decomposed at relatively low temperatures. Only the triphenylphosphine sulfide of CuCl, [CuCl(SPPh₃)₃], was considered sufficiently robust.

Table II. Hydrolysis Tests for Copper(I) Phosphite Complexes

Complex	Time to Wetness & Clumping Behavior @ 40 °C, 80% Relative Humidity
CuCl(P(OPh)$_3$)	2 hours
CuBr(P(OPh)$_3$)	6 hours
CuCl(BPP)	>28 days
CuBr(BPP)	>28 days
CuCl(MBPOP)	>28 days
CuBr(MBPOP)	>28 days
(CuCl)$_2$(BHTPD)	>28 days
(CuBr)$_2$(BHTPD)	>28 days
CuCl(CPD)	>28 days
CuBr(CPD)	>28 days

Chart 3

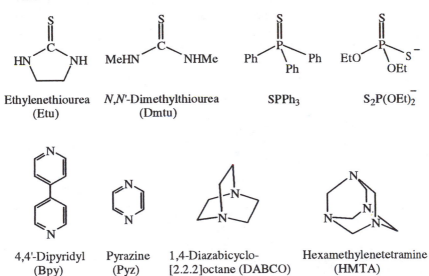

| Ethylenethiourea (Etu) | N,N'-Dimethylthiourea (Dmtu) | SPPh$_3$ | S$_2$P(OEt)$_2^-$ |

| 4,4'-Dipyridyl (Bpy) | Pyrazine (Pyz) | 1,4-Diazabicyclo-[2.2.2]octane (DABCO) | Hexamethylenetetramine (HMTA) |

The Cu(I) products with bridging nitrogen ligands, only a few of which are listed (Bpy, Pyz, DABCO, and HMTA), form inorganic networks *(16,17)* and might be expected to show fairly high thermal stability as a result. However, this was not the case. Moreover, the complexes that incorporated conjugated ligands (e. g., Bpy and Pyz) were colored because of metal-to-ligand charge transfer.

276

Table III. Copper(I) Sulfur and Nitrogen Complexes Prepared

Complex	Color	Cu, Theory	Cu, Expt.	Dec. Point
CuBr(Etu)	white	25.9%	26.1%	160 °C
CuCl(Dmtu)$_2$	white	20.7	22.0	160
CuCl(SPPh$_3$)$_3$	white	6.5	6.5	225
Cu(S$_2$P(OEt)$_2$)	white	25.5	25.3	187
Cu(S$_2$P(OEt)$_2$)(SPPh$_3$)$_2$	white	7.6	6.8	180
CuBr(Bpy)	red	21.2	20.8	215
CuCl(Pyz)	red	35.5	36.4	125
(CuCl)$_2$(DABCO)	lt. brown	41.0	40.0	161
(CuCl)$_2$(HMTA)	sandy	37.6	37.4	180

Copper(II) Salts

Copper(II) compounds have the disadvantage, for their use as polymer additives, of being colored. However, simple anhydrous salts of Cu(II) do not usually have high molar absorptivities. In combination with oxidizable ligands, thermal reductive elimination reactions, such as that shown in equation 3, can occur and thereby form low-valent copper.

A series of copper(II) salts with oxidizable ligands was synthesized. The resultant compounds are listed in Table IV. Anions derived from both

Table IV. Copper(II) Salts Prepared

Complex	Color	Cu, Theory	Cu, Expt.	Dec. Point
Cu(HPO$_3$)•1.5 H$_2$O	pale blue	37.3%	37.5%	126 °C
Cu(PhPO$_3$)•1.5 H$_2$O	pale blue	25.8	26.2	344
Cu(H$_2$PO$_2$)$_2$	v. pale blue	32.8	33.0	55[a]
Cu(HPhPO$_2$)$_2$	v. pale blue	18.4	18.1	99
Cu(Ph$_2$PO$_2$)$_2$	pale blue-violet	12.8	13.3	299
Cu(C$_2$O$_4$)	pale blue	41.4	41.0	300
Cu(O$_2$CH)$_2$•1.5 H$_2$O	pale blue	35.2	34.7	199

[a]Decomposes over several days at 25 °C.

phosphonic acid (H$_3$PO$_3$, P(III)) and phosphinic acid (H$_2$PO$_2$, P(I)) were chosen, since these low oxidation states are subject to oxidation, which produces salts of phosphoric acid (H$_3$PO$_4$, P(V)). It can readily be observed that replacement of the nonacidic P–H with P–Ph conferred significant stability enhancement to the resulting salt. Thus, the phenylphosphonate and diphenyl-phosphinate salts are the most promising in terms of stability. The previously reported (4) oxalate and formate salts are also listed in Table IV for comparison.

Prospects for PVC Additives

Many of the compounds listed in Tables I, III, and IV have been tested for their promotion of the thermal cross-linking of PVC (as judged by gel yield). The desired behavior for a promising additive candidate is the formation of a high gel yield without significant sample mass loss during the pyrolysis (4). Some results are listed in Table V.

Table V. Gelation and Mass Loss Results

Additive Complex (10% by mass added to PVC)	Number of Trials	PVC Gel Yield, 1 h @ 190 °C	PVC Mass Loss During Pyrolysis, 1 h @ 190 °C
Control	2	<5%	0.3%
CuCl(P(OPh)$_3$)	2	82	4.7
CuBr(P(OPh)$_3$)	2	69	2.1
CuCl(BPP)	3	69	0.6
CuBr(BPP)	4	50	1.2
CuCl(MBPOP)	3	39	3.6
CuBr(MBPOP)	4	45	2.1
(CuCl)$_2$(BHTPD)	4	62	2.6
(CuBr)$_2$(BHTPD)	4	63	0.8
CuCl(CPD)	4	68	2.4
CuBr(CPD)	3	45	1.6
CuCl(SPPh$_3$)$_3$	2	65	1.5
Cu(S$_2$P(OEt)$_2$)	2	<5	3.3
Cu(S$_2$P(OEt)$_2$)(SPPh$_3$)$_2$	2	<5	1.6
CuBr(Bpy)	2	82[a]	0.5[a]
(CuCl)$_2$(HMTA)	2	82[a]	5.1[a]
Cu(PhPO$_3$)•1.5 H$_2$O	4	69	1.9
Cu(Ph$_2$PO$_2$)$_2$	3	86	1.2
Cu(C$_2$O$_4$)	2	50	4.6
Cu(O$_2$CH)$_2$•1.5 H$_2$O	2	65	3.9

[a]Results at 200 °C.

Several conclusions may be drawn from the data in this table. None of the copper-based additives appears to promote significant dehydrochlorination of PVC at 190 °C. These observations stand in marked contrast to the behavior seen for more Lewis-acidic iron chlorides or copper(I) complexes of the more basic PPh$_3$ (4). Many of the new additives produce moderate-to-large amounts of PVC gelation. It should be noted that this process is occurring well below the observed decomposition temperatures of the additives. In fact, the only tested complexes producing relatively little gelation, those of diethyl dithiophosphate, had decomposition points quite close to the testing temperature of 190 °C. Hence, it can now be concluded that extensive decomposition of the additive compound is not always needed for the promotion of significant polymer gelation.

Particularly promising were the new hindered phosphite complexes of Cu(I). Phosphites are widely used as antioxidant stabilizers for PVC and other polymers (20–22). Thus their incorporation into new smoke-suppressant additives is a natural step. Comparison of the P(OPh)$_3$ complexes to those of the new phosphites reveals similarly favorable gel yields. This result is noteworthy, since the use of the increasingly larger monophosphites decreases the copper content of the additives. The exceptions to this trend are the complexes of BHTPD, which coordinate two metals atoms per ligand molecule and thus have a relatively high copper content (see Table I). Of course, the important advantage imparted by the use of the hindered phosphites is their vastly increased hydrolytic stability (see Table II).

Although the ideal polymer additive (aside from colorants) is colorless, several copper(II) complexes: Cu(PhPO$_3$), Cu(Ph$_2$PO$_2$)$_2$, and Cu(C$_2$O$_4$), may have promise as smoke suppressants. These species are very stable thermally and yet promote good-to-excellent gelation at temperatures well below their decomposition points. Moreover, since they are relatively simple salts of Cu(II), they are expected to be inexpensive, especially when compared to the more exotic organophosphite complexes.

Conclusions and Future Prospects

A new class of copper(I) phosphite complexes, having enhanced hydrolytic stability, has been identified. Initial tests of these compounds reveal good cross-linking promotion of PVC at elevated temperatures without evidence of PVC dehydrochlorination. Similarly promising results have been obtained for relatively simple salts of copper(II) bearing oxidizable ligands such as oxalate, phenylphosphonate, and diphenylphosphinate.

We are continuing our investigations in this area. It is our expectation that smoke and flame tests will further demonstrate the advantages of low-valent metal-based additives used in place of, or in combination with, traditional Lewis-acidic additives, in order to address the problems of smoke and flame from PVC.

Acknowledgment

We thank the National Science Foundation (Grant No. CHE-9983374) and the Camille and Henry Dreyfus Foundation (Grant No. TH-99-010) for partial support of this work. We also thank Dover Chemical Co. and Amfine Chemical Co. for providing samples of organophosphites.

References

1. Starnes, W. H., Jr.; Edelson, D. *Macromolecules* **1979**, *12*, 797–802.
2. Lattimer, R. P.; Kroenke, W. J. *J. Appl. Polym. Sci.* **1980**, *25*, 101–110.
3. Starnes, W. H., Jr.; Girois, S. *Polym. Yearbook* **1995**, *12*, 105–131.
4. Pike, R. D.; Starnes, W. H., Jr.; Jeng, J. P.; Bryant, W. S.; Kourtesis, P.; Adams, C. W.; Bunge, S. D.; Kang, Y. M.; Kim, A. S.; Kim, J. H.; Macko, J. A.; O'Brien, C. P. *Macromolecules* **1997**, *30*, 6957–6965, and references cited therein.
5. Starnes, W. H., Jr.; Wescott, L. D., Jr.; Reents, W. D., Jr.; Cais, R. E.; Villacorta, G. M.; Plitz, I. M.; Anthony, L. J. In *Polymer Additives*; Kresta, J. E., Ed.; Plenum: New York, 1984; pp 237–248.
6. Wescott, L. D., Jr.; Starnes, W. H., Jr.; Mujsce, A. M.; Linxwiler, P. A. *J. Anal. Appl. Pyrol.* **1985**, *8*, 163–172.
7. Ebert, G. W.; Rieke, R. D. *J. Org. Chem.* **1988**, *53*, 4482–4488.
8. Ginah, F. O.; Donovan, T. A., Jr.; Suchan, S. D.; Pfennig, G. R.; Ebert, G. W. *J. Org. Chem.* **1990**, *55*, 584–589.
9. Yanagisawa, A.; Hibino, H.; Habaue, S.; Hisada, Y.; Yamamoto, H. *J. Org. Chem.* **1992**, *57*, 6386–6387.
10. Corey, E. J.; Semmelhack, M. F. *J. Am. Chem. Soc.* **1967**, *89*, 2755–2757.
11. Kawaki, T.; Hashimoto, H. *Bull. Chem. Soc. Jpn.* **1972**, *45*, 3130–3132.
12. Nakanishi, S.; Oda, T.; Ueda, T.; Otsuji, Y. *Chem. Lett.* **1978**, 1309–1312.
13. Yamada, Y.; Momose, D. *Chem. Lett.* **1981**, 1277–1278.
14. Nishizawa, Y. *Bull. Chem. Soc. Jpn.* **1961**, *34*, 1170–1178.
15. Pike, R. D.; Starnes, W. H., Jr.; Carpenter, G. B. *Acta Crystallogr., Sect. C* **1999**, *55*, 162–165.

16. Graham, P. M.; Pike, R. D.; Sabat, M.; Bailey, R. D.; Pennington, W. T. *Inorg. Chem.*, **2000**, *39*, 5121–5132.
17. Pike, R. D.; Starnes, W. H., Jr.; Graham, P. M. In *Polymer Additives*; Al-Malaika, S., Golovoy, A., Wilkie, C. A., Eds.; Blackwell Science: Oxford, U.K., 2001, in press.
18. Drew, M. G. B.; Forsyth, G. A.; Hasan, M.; Hobson, R. J.; Rice, D. A. *J. Chem. Soc., Dalton Trans.* **1987**, 1027–1033.
19. Tiethof, J. A.; Hetey, A. T.; Meek, D. W. *Inorg. Chem.* **1974**, *13*, 2505–2509. These authors claim to form [CuCl(SPPh$_3$)], rather than our observed product, [CuCl(SPPh$_3$)$_3$].
20. Stevenson, D.; Stein, D. In *Chemistry and Technology of Polymer Additives*; Al-Malaika, S., Golovoy, A., Wilkie, C. A., Eds.; Blackwell Science: Oxford, U.K., 1999; Chapter 6.
21. Troitzsch, H. J. In *Plastics Additives Handbook*, 4th ed.; Gächter, R., Müller, H., Eds.; Hanser/Gardner: Cincinnati, OH, 1993; Chapter 12.
22. Gugumus, F. In *Plastics Additives Handbook*, 4th ed.; Gächter, R., Müller, H., Eds.; Hanser/Gardner: Cincinnati, OH, 1993; Chapter 1.

Chapter 22

Processing and Thermo-Oxidative Aging of High-Impact Polystyrene-Containing Fire Retardant

Kenneth Moller, Jukka Lausmaa, and Antal Boldizar

Department of Chemistry and Materials Technology, SP Swedish National Testing and Research Institute, Box 857, S-501 15 Boras, Sweden

High Impact Polystyrene (HIPS) with and without a fire retardant, deca-bromodiphenylether (deca-BDE), has been subjected to an accelerated thermo-oxidative aging corresponding to a normal in-door service time of about ten years. The purpose of the study was to examine if HIPS with deca-BDE can be recycled without loss of important properties. Test specimens of HIPS were aged at 80 °C for 1400 hours. Changes in mechanical properties, e.g. elongation at break, were measured. The material with deca-BDE added was found to retain its mechanical properties. Inductive Coupled Plasma Mass Spectrometry (ICP-MS) was used to examine if deca-BDE leaves the HIPS material during processing and aging. The ICP investigation showed that no significant amount of deca-BDE has left the HIPS material. Matrix Assisted Laser Desorption/Ionization Mass Spectrometry was used to investigate changes of deca-BDE during processing and aging. No evidence of degradation was found. In this investigation nothing has appeared indicating that HIPS containing deca-BDE is not well suited for recycling.

Introduction

The need for fire resistant polymeric materials is continuously growing. This is partly due to the increasing use of for example, electronic equipment and polymeric building materials and partly caused by more stringent flammability requirements.

Decabromodiphenylether (deca-BDE) is a widely used, high performance fire retardant (FR) for thermoplastics. It also has a low toxicity and bioaccumulation (1). In Europe in recent years, there has been a general desire to restrict the use of halogen containing compounds partly based on perceived recycling difficulties.

One way of reducing the displacement of halogen containing materials in the environment or decreasing the need for costly destruction of such materials is recycling. In order to be accepted as construction materials, however, mechanical, durability, aesthetical, and other properties must be retained in the recycled materials.

One can, however, suspect that deca-BDE is not fully stable during aging and processing. Instead, from a durability point of view harmful degradation products like HBr might be formed. It is well known that an autocatalytic formation of HCl causes serious degradation of polyvinylchloride (PVC) (2). The suspected instability in combination with the high concentration of deca-BDE normally used in HIPS might seriously restrict the possibility for recycling.

The purpose of this study was, therefore, to examine if HIPS with FR (FR-HIPS) in comparison with HIPS without FR (NFR-HIPS) can be recycled without serious loss of important properties, including fire performance. HIPS with and without deca-BDE has, for that reason, been subjected to accelerated thermo-oxidative aging corresponding to a normal in-door service time of ten years and analyzed with respect to mechanical properties as well as FR retainment.

Other additives like antimony oxide were not considered to be unstable in the same way as deca-BDE and should not affect the recycling potential of FR-HIPS in a negative way.

Experimental Description

Problems and Solution

The conclusion, based on initial Differential Scanning Calorimetry (DSC), was that we could try thermo-oxidative aging at 95 °C, although the position of the glass transition region was difficult to evaluate. After a short period of aging, it became obvious that the aging temperature was too high, resulting in shrinkage and distortion of samples. A more thorough examination with DSC indicated the glass transition region to be between 87 and 95 °C. A second aging series was started at 80 °C, in order to age well below the glass transition region. If aging was performed at temperatures in the transition region or above, we would most probably introduce irrelevant degradation processes.

Since the aging temperature was reduced, the aging time should be increased in order to obtain the desired aging effect. The acceleration factor assumed for the thermo-oxidative aging corresponds roughly to a doubling of the aging rate for each increase of 10°C. For a simulated service time of ten years, this would lead to an aging time of about 1400 hours at 80°C for our samples. The estimation above is based on the assumption that the rubber phase in impact modified polystyrene is the most vulnerable one (3,4). According to Dixon, it is recommended to use a relatively low activation energy (60-70 kJ/mol) when calculating the acceleration factor in order not to overestimate the expected service life(5). Most likely the expected service life will be closer to twenty years then the ten years at which this study is aimed.

Materials and Sample Preparation

The HIPS materials were delivered by Nova Chemicals. The stabilizer packages for both grades were normal for HIPS, according to the supplier.

Both materials were extruded to about 0.2 mm thin strips by using a single screw Brabender PLE 651 extruder. The temperature profile in the extruder was 160, 180, 200, 200, and 210°C and the speed of rotation of the screw was 60 rpm. The temperature profile used is recommended by the HIPS producer and has been verified by BASF and Dow.

Thicker test specimens were also manufactured by injection molding according to ISO 294 and test specimen type 1B. The following parameters were used: melt temperature 230°C, injection pressure 1000 bar, and a follow-up pressure of 700 bar. These parameters are recommended for virgin HIPS.

The reason for using two different sample thicknesses is that the thick samples correspond to dimensions of normal FR containing HIPS products, while in thin samples thermo-oxidative degradation is not as restricted by oxygen diffusion as for thick samples. Consequently, degradation will be more severe in the thin samples. The degradation of a thin sample can also be regarded as corresponding to the degradation of a thin surface region of a thick sample. Moreover, loss of additives depends on the relation between exposed surface area and bulk volume. The thick samples were used for performing UL 94 testing to establish the ignition behavior of the FR and NFR-HIPS material, both before and after aging and recycling.

Aging

Aging of HIPS and HIPS-FR took place in 0.3 m³ Salvis heating cabinets. The temperature was 80 °C and the duration of aging was 1400 hours corresponding to about ten years of normal room temperature service. About 50 test specimens of each type were aged.

Tensile and Impact Strength Testing

An Instron 5566 tensile tester equipped with a non-contacting Instron 2663 video extensiometer was used to acquire data for tensile stress and elongation at break according to ISO 527. Crosshead speed was 20 mm/min. The impact strength test was performed according to ISO 179 using a Zwick Impact Strength Tester. For each test above at least ten specimens were used. Mean values and standard deviations were calculated.

Determination of Total Bromine Content

A Micromass Platform Inductive Coupled Plasma-Mass Spectrometer (IC - MS) was used to determine the total bromine content in extruded FR-HIPS before and after aging.

Ground FR-HIPS was mixed with potassium hydroxide, hydrazine sulfate, and benzoic acid. The mixture was burned in a bomb calorimeter in oxygen in order to convert all bromine compounds into bromide, which was subsequently dissolved in water. The solution was injected into the ICP instrument for total bromine content determination. Both unaged and aged materials were investigated.

Degradation of deca-BDE

In order to investigate if deca-BDE degrades due to aging and/or processing, pure deca-BDE and deca-BDE removed from unaged and aged FR-HIPS were examined with Matrix Assisted Laser Desorption and Ionization - Mass Spectrometry (MALDI-MS).

The MALDI instrument used was a Bruker BIFLEX™III. The 337 nm nitrogen UV laser has an intensity of 107 Wcm^{-2} and a pulse length of 5 ns at ~ 1 Hz. The mass resolution in the time-of-flight mass spectrometer is >10,000.

FR was removed from HIPS by a procedure involving dissolution of the plastic in tetrahydrofuran (THF) and filtering of the solution in order to remove carbon black. The polymer and the fire retardant were separated by gel permeation chromatography (GPC). The fraction containing the FR was collected and the solvent was evaporated. For the MALDI analysis THF was used as solvent and picolinic acid (PA) as matrix. Irganox 1010 and Irganox 1076 were added and used for mass calibration.

Fire Classification

In order to determine whether the material being tested passed the required ignitability behavior both before and after aging and recycling, injection molded samples were subjected to classification according to UL 94.

The NFR-HIPS samples were classified according to the "horizontal burn" characterization in the UL 94 standard while the FR-HIPS samples were subjected to classification according to the "vertical burn" classification described in UL94. The results are summarized in the next section.

Results

Mechanical Testing

The elongation at break for both the aged and unaged extruded FR- and NFR-HIPS is shown in Figure 1.

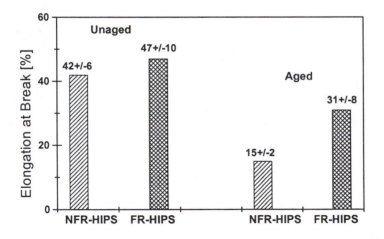

Figure 1: Elongation at break for extruded FR-HIPS and NFR-HIPS before and after aging.

The values for both NFR-HIPS and FR-HIPS have decreased after thermo-oxidative aging. The decrease after aging is, however, more pronounced for the NFR-HIPS.

286

The corresponding results for injection molded HIPS are shown in Figure 2. There is only a minor change due to aging for the FR-HIPS, while the NFR-HIPS shows a very distinct reduction in elongation at break, i.e., greater than 50%.

The explanation for the difference in behavior between extruded and injection molded test specimens (Figures 1 and 2) is probably the restriction in oxygen diffusion for thicker injection molded samples, which reduces thermo-oxidative degradation inside the samples. Most probably the FR-HIPS samples contain a higher % of stabilizer than the NFR-HIPS samples. The material used in this study is commercial grade and thus exact information concerning the composition is not available.

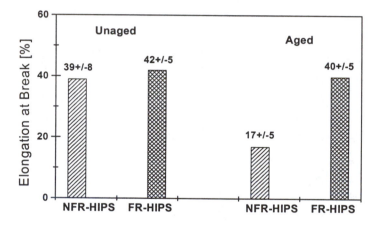

Figure 2: Elongation at break for injection molded FR-HIPS and NFR-HIPS before and after aging.

Although elongation at break is regarded to be more sensitive to aging of polymeric materials than tensile stress, the latter is also of interest. In many cases stress at break increases by aging. This is generally due to an increase in cross-linking of the polymer as a part of the aging process (6).

The value of stress at break for extruded test specimens is shown in Figure 3 Comparing the results for FR-HIPS one finds a very small change between aged and unaged samples. For NFR-HIPS, on the other hand, there is a significant increase in stress at break indicating that some kind of degradation has taken place as a result of aging. Again, this is probably due to a higher portion of stabilizers in the FR-HIPS than in the NFR-HIPS.

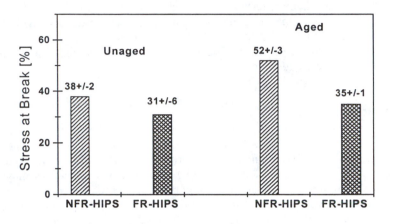

Figure 3: Stress at break for injection molded FR-HIPS and NFR-HIPS before and after aging.

Results for the impact strength measurements regarding injection molded HIPS are shown in Figure 4. The result is quite apparent. No significant change is found for FR-HIPS, while a 50% reduction has occurred for NFR-HIPS.

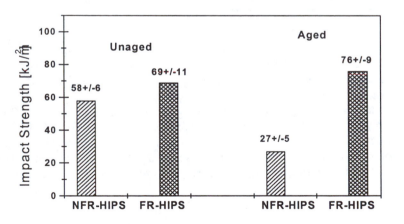

Figure 4: Impact strength for injection molded FR-HIPS and NFR-HIPS before and after aging.

Thus, all mechanical testing performed indicate that the FR-HIPS used in this study withstands thermo-oxidative aging much better than does NFR-HIPS. It should, however, be emphasized that the reason for this result is not necessarily due to the presence of deca-BDE. As explained above this effect is probably due to a co-additive.

Total Bromine Content

The total bromine content was determined by ICP-MS before and after aging for extruded HIPS. Thin extruded films were investigated as they were subjected to more severe degradation compared to the thicker injection molded samples. Moreover, the thin extruded films have a larger exposed surface area relative to the bulk volume compared to the thicker injection molded samples. Large surface area relative to bulk volume can cause higher emission of additives. Thus, these samples are a more severe test of the retention of the fire retardant.

Table I. Total bromine content in unaged and aged FR-HIPS.

	^{79}Br	^{81}Br
Aged	11.1	11.1
Unaged	11.1	11.1

As is seen in Table I, the bromine content is 11 mass% of deca-BDE in unaged and aged FR-HIPS. Both isotopes, ^{79}Br and ^{81}Br, were used in the determination and they gave the same results. Thus, no significant loss of deca-BDE was observed.

It must be pointed out that this investigation was focused on the durability of FR-HIPS material, i.e., on possible significant losses of deca-BDE which might affect the fire retarding properties. Although very important from an environmental point of view, possible losses of small amounts of deca-BDE to the environment during ageing or processing were outside of the scope of this investigation.

Degradation of deca-BDE

Ionization in MALDI is regarded as very "soft", i.e., only molecular ions are formed of analyte molecules. In the case of deca-BDE, however, there is a significant debromination in the ionization process. This effect is also seen in negative ion spectra where Br^-, Br_2^-, and Br_3^- ions are detected. The debromination occurs both for pure deca-BDE and for deca-BDE extracted from unaged and aged FR-HIPS.

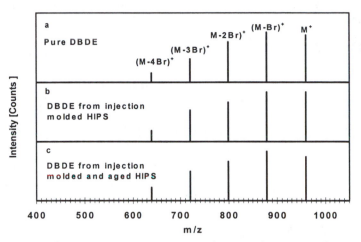

Figure 6a-c: Schematic MALDI MS-spectra for pure deca-BDE, deca-BDE from unaged FR-HIPS, and deca-BDE from aged FR-HIPS, respectively.

Comparing the spectra in Figure 6 one finds no significant differences, indicating that degradation of deca-BDE, other than that caused by the ionization process, does not take place. Hamm et. al. have investigated HIPS containing deca-BDE after repeated processing of the material. They found no degradation of the fire retardant (7).

Fire Classification

The results of the fire classification of both aged and unaged FR- and NFR-HIPS indicated that the NFR-HIPS did not retain its HB rating after aging while the FR-HIPS retained its V0 rating. The results are summarized in Table II.

Table II. Results of UL 94 classification of FR- and NFR-HIPS strips

	FR-HIPS	NFR-HIPS
Aged	V0	Failed HB
Unaged	V0	HB

Conclusions

Mechanical tests clearly show that HIPS containing deca-BDE has improved aging properties compared to the HIPS used in this study, without deca-BDE. In fact, the investigation indicates that unaged FR-HIPS has slightly better mechanical properties than unaged NFR-HIPS.

There is no indication of loss of deca-BDE due to aging. No degradation of deca-BDE caused by aging was found. Additional work should, however, be done in order to confirm this result.

Importantly, the fire retardant behavior of the FR-HIPS is retained after aging. Thus, the FR-HIPS is able to pass its fire retardant properties onto a new generation of products.

One can conclude, that in this investigation nothing has appeared indicating that HIPS containing deca-BDE is not well suited for recycling.

References

1. Norris, J.M.; Ehrmantraut, J.W.; Gibbons, C.L.; Kociba, R.J.; Schwetz, B.A.; Rose, J.Q., Humiston C.G.; Jewett, G.L.; Crummett, W.B.; Gehring, P.J.; Tirsell, J.B., Brosier, J.S. *Appl. Polym. Sym.* **1973**, *22*, 195-219.
2. Salman, R.S.; Al-Shama'a, N.D. *Polym.-Plast. Technol. Eng.* **1991**, *30(4)*, 343-349.
3. Shimada, J.; Kabuki, K. *J. Appl. Polym. Sci.* **1988**, *12*, 671-685.
4. Kelen, T., *Polymer Degradation*, Van Nostrand Reinhold Company, New York, 1983.
5. Dixon, R.K., *IEEE. Transaction on Elec. Insul.* **1980**, *E1-15, No. 4*, 331-340.
6. Grassie, N.; Scott, G. *Polymer Degradation and Stabilisation*, Chapter 4, Cambridge University Press, 1985.
7. Hamm, S., Schulte, J., Strikkeling, M., Maulshagen, A., Poster at Dioxin 99, Venice, 1999.

Assessment and Performance

Organic polymers are combustible. They decompose thermally, decomposition products burn, smoke is generated, and products of combustion are generated that are highly toxic, even if only CO and CO_2. Fire performance combines thermal decomposition, ignition, flame spread, heat release, ease of extinction, smoke obscuration, toxicity, and other properties. Assessment of materials for research purposes or for regulation may require different assessment tools. Regulation requires that set of properties relevant to the end application. Research requires that set of properties needed to understand the mechanism of a material's performance at the level of a specific investigation. The first paper (Hirschler) in this section discusses polymer breakdown mechanisms, fire properties, and their use and application in fire modeling and fire hazard assessment. In the second paper (Price et al.), three research techniques for studying polymer combustion at a given laboratory are illustrated with specific material examples. In the third paper (Metcalfe and Tetteh), FTIR is used in the laboratory to identify toxic gases from burning polymers assisted with a chemometric methodology. In the remaining three papers, three different polymer systems—two synthetic (a thermoplastic and a thermoset) and one natural (wood)—are evaluated using available assessment tools. These three papers taken together provide a tutorial on what can be learned about material performance using available methodology.

Chapter 23

Fire Performance of Organic Polymers, Thermal Decomposition, and Chemical Composition

Marcelo M. Hirschler

GBH International, 2 Friar's Lane, Mill Valley, CA 94941

Organic polymers are combustible: they decompose thermally, and decomposition products burn. Polymer breakdown rates and mechanisms vary broadly, as does the subsequent combustion. A combustion process is self-sustaining if burning gases from polymer thermal decomposition feed back sufficient heat to the condensed phase to continue producing fuel vapors. Heat transferred to the polymer causes generation of flammable volatiles, which react with atmospheric oxygen to generate heat, part of which is transferred back to the polymer. The fuel generated by a burning organic polymer is related to the original polymer structure and depends on the reaction of the polymer or its breakdown products with the oxidant before entering the flame zone. Polymer fire characteristics often differ from those of traditional materials they are replacing. Fire safety combines: thermal decomposition, ignition, flame spread, heat release, smoke obscuration, and other properties. Overall fire safety is achieved if products meet certain objectives, by combining various properties and calculating results based on fire models. This work addresses how to assess fire safety, and effects of polymeric fuel chemical composition.

Introduction

Organic polymers are combustible: if sufficient heat is supplied they thermally decompose, and the products burn. Combustion will be self-sustaining if the burning gases return sufficient heat to the polymer condensed phase to continue fuel vapor production, and cause continuous feedback loop if the material continues burning. Heat transferred to the polymer generates flammable volatiles, which then react with the oxidant in the atmosphere around the condensed phase to generate heat. A part of this heat is transferred back to the polymer to continue the combustion process. The nature of the fuel supplied to the flame when an organic polymer burns depends on the structure of the original polymer and on the extent to which the polymer or its breakdown products react with oxidants before entering the flame zone. The chemistry of most flames, including those associated with burning polymers, is similar, involving largely the same chain-propagating radicals.

Chemical Classification of Polymers by their Molecular Structure

The most useful classification method for polymers *(1)* is based on their chemical composition, which gives an indication polymer reactivity including thermal decomposition and fire performance. Organic polymers can be divided based on their chains and substituents: carbon-carbon bond chains without heteroatoms, carbon chains with either oxygen, nitrogen, chlorine or fluorine atoms as heteroatoms, and non carbon chains.

Generic Chemical Breakdown Mechanisms

Four classes of chemical mechanisms are important in polymer thermal decomposition of addition polymers: random-chain scission, end-chain scission, chain-stripping and cross-linking. Thermal decomposition of a polymer generally involves more than one of these classes of reactions, but these general classes provide a useful conceptual framework. Chain-scission mechanisms involve atoms in the main polymer chain while both chain stripping and cross-linking involve principally side chains or groups. Although decomposition of some polymers follows a single mechanism, many cases involve combinations of mechanisms.

Chain scission: random- and end-initiated: Among simple thermoplastics, the most common reaction mechanism is bond breaking in the main polymer chain. Chain scissions may occur randomly (random-chain scission) or at the chain end (end-chain scission). End-scission results in monomer production, and is often known as 'unzipping.' Random-scission results in generation of both monomers and oligomers as well as a variety of other chemical species. The type and distribution of volatile products depend on the relative volatility of the resulting molecules. Chain scission is common in vinyl polymers: the process is a multi step radical chain reaction with all the general features of such reaction mechanisms:

initiation, propagation, branching, and termination steps, with initiation producing the free radicals. Random scissions involve the breaking of a main chain bond at random location, all such main chain bonds being equal in strength. End-chain scissions involve breaking off of a small unit or group at the end of the chain: a monomer unit or a smaller substituent. Vinyl polymers are derived from a vinyl repeating unit, namely - $[CH_2 -CH_2]_n$ -, with n the number of repeating monomers. If hydrogen atoms are substituted the repeating unit is: - $[C (W X) - C (Y Z)]_n$ -, where W, X, Y, and Z are substituents. In order for hydrogen to be transferred from the chain to the radical site, it must pass around either Y or Z. If Y and Z are hydrogen, this is not difficult due to their small size. However, if larger substituents are bound to the alpha carbon (i.e., Y and Z are larger groups), hydrogen transfer to the radical site is more difficult, due to 'steric hindrance.' Polymers with generally large Y and Z substituents result in a high monomer yield, characteristic of unzipping reactions, while polymers where Y and Z are small, form small amounts of monomer since other mechanisms dominate. Table 1 shows the monomer yield for various common vinyl polymers. In general most polymers undergoing chain scission release small flammable vapors and will have poor fire performance: the burning vapors will easily ignite the solid.

Chain stripping: Chain stripping involves the loss of a small molecule by side-chain substituent release from the main chain, forming small molecules. In elimination reactions, the bonds connecting side groups of the polymer chain to the chain itself are broken, with the side groups often reacting with other eliminated side groups. The reactions generally yield products small enough to be volatile. Two adjacent side groups react, in cyclization reactions, to form bonds, and develop a cyclic structure. The process is also important in char formation, as the residue is richer in carbon than the original polymer. Polymers undergoing chain stripping can have poor fire performance if the molecules lost are flammable and the stripped chain does not form char, but will tend to have excellent fire performance if the molecule released is a halogen acid, which will act as a vapor phase free radical scavenger.

Cross-linking: Cross-linking partially involves the main chain. It generally occurs after substituent stripping and involves bond creation between adjacent polymer chains. The process is critical in char formation, as it generates a more compact structure, which is then less likely to lead to volatiles. Cross-linking also increases melting temperature and can render a material infusible. While char formation is chemical, its significance is largely due to physical properties. Clearly, char left in the solid phase means that less flammable gases will be released on thermal decomposition. Moreover, char is also a barrier between the heat source and the virgin polymeric material, so that heat flow to the virgin material is reduced as the char layer

Table 1 - Monomer Yield On Thermal Decomposition of Addition Vinyl Polymers

Polymer	W	X	Y	Z	Yield (wt%)	Decomp. Mechan. *
Poly(methyl methacrylate)	H	H	CH_3	CO_2CH_3	91-98	E
Polymethacrylonitrile	H	H	CH_3	CN	90	E
Poly(α methyl styrene)	H	H	CH_3	C_6CH_5	95	E
Polyoxymethylene #	-	-	-	-	100	E
Polytetrafluoroethylene	F	F	F	F	95	E
Poly (methyl atropate)	H	H	C_6H_5	CO_2CH_3	>99	E
Poly (p bromo styrene) §	H	H	H	C_6H_4Br	91-93	E
Poly (p chloro styrene) §	H	H	H	C_6H_4Cl	82-94	E
Poly (p methoxy styrene) §	H	H	H	C_7H_7O	84-97	E
Poly (p methylstyrene) §	H	H	H	C_7H_7	82-94	E
Poly (α deuterostyrene)	H	H	D	C_6H_5	70	E
Poly (α, β, β trifluorostyrene)	F	F	F	C_6H_5	44	E/R
Polystyrene	H	H	H	C_6H_5	42-45	E/R
Poly (m methyl styrene)	H	H	H	C_7H_8	44	E/R
Poly (β deutero styrene)	H	D	H	C_6H_5	42	E/R
Poly (β methyl styrene)	H	CH_3	H	C_6H_5		E/R
Poly (p methoxy styrene) ¶	H	H	H	C_7H_7O	36-40	E/R
Polyisobutene	H	H	CH_3	CH_3	18-25	E/R
Polychlorotrifluoroethylene	F	F	Cl	F	18	E/S
Poly (ethylene oxide) #	-	-	-	-	4	R/E
Poly (propylene oxide) #	-	-	-	-	4	R/E
Poly (4 methyl pent-1-ene)	H	H	H	C_4H_9	2	R/E
Polyethylene	H	H	H	H	0.03	R
Polypropylene	H	H	H	CH_3	0.17	R
Poly (methyl acrylate)	H	H	H	CO_2H_3	0.7	R
Polytrifluoroethylene	F	F	H	F	-	R
Polybutadiene #	-	-	-	-	1	R
Polyisoprene #	-	-	-	-	0-0.07	S
Poly (vinyl chloride)	H	H	H	Cℓ	-	S
Poly (vinylidene chloride)	H	H	Cℓ	Cℓ	-	S
Poly (vinylidene fluoride)	H	H	F	F	-	S
Poly (vinyl fluoride)	H	H	H	F	-	S
Poly (vinyl alcohol)	H	H	H	OH	-	S
Polyacrylonitrile	H	H	H	CN	5	C

Notes: *:Decomposition Mechanism: R: Random chain scission, E: End chain scission; S: chain stripping; C: cross-linking. #: Not of general formula [CWX-CYZ]n . §: Cationic polymerization. ¶: Free-radical polymerization reaction occurs immediately after an initiation reaction, no unzipping.
Table source: Reference *(1)* - Copyright: Marcelo M. Hirschler.

thickens. The decomposition rate is consequently reduced, depending on char properties. If burning of the volatiles is the heat source, not only will the fraction of the incident heat flux flowing to the material be reduced, but the overall incident heat flux will be reduced. Char formation is usually, but unfortunately not always, advantageous. Solid-phase char combustion can cause sustained smoldering combustion. Thus, by enhancing the charring tendency of a material, flaming combustion rates may be reduced, but perhaps at the expense of creating a source of smoldering combustion that would not otherwise have existed. The opposite is also true: inhibition of smoldering combustion can enhance flaming combustion.

Exothermic reactions and self-heating: Self-heating of polymers is a serious problem during production and storage of some materials, typically cellulosic materials. The most common example (usually considered the only case) is cotton cloth impregnated with linseed oil. The problem is, however, much broader than cellulosic polymers and can be found with synthetic condensation polymers, typically elastomeric.

Implications of thermal decomposition to fire properties

Thermal decomposition of polymers is often studied because of its importance in terms of fire performance. Early on, it was shown that, for many polymers, the limiting oxygen index (LOI, an early flammability measure) could be linearly related to char yield as measured by TGA (2). Then, since char yield can be computed from structural parameters, LOI should be computable, and for pure polymers having substantial char yields, it is fairly computable. However, neither minimum decomposition temperature nor 1% thermal decomposition temperature correlates with LOI. Although low flammability is usually associated with high minimum thermal decomposition temperature, there is no easy way to compare the two, with notable examples of polymers with low thermal stability and low flammability (3-4). This approach has fallen into disrepute, in view of the lack of confidence today in the LOI technique. Recent work shows that micro calorimeter thermoanalytical data can predict some heat release information, but only on pure polymers (5). Mechanisms of action of fire retardants and potential effectiveness of fire retardants can be well predicted from thermal decomposition activity, particularly if an added understanding exists of the chemical reactions involved (6). The newest approach is mass loss calorimetry (actually mass loss and ignitability data only with a cone heater: ASTM E 2102), using heat fluxes typical of fires. Independent of the understanding on how to predict fire performance from thermal decomposition data, this is a critically achievable goal, as polymers cannot burn if they do not break down (1, 3).

Fire Performance of Polymers

Fire hazard is the result of a combination of factors including ignitability, ease of extinction, flammability of the products generated, amount of heat released on burning, rate of heat release, flame spread, smoke obscuration and smoke toxicity, as well as the fire scenario. Compliance with fire properties is often required by different codes or regulations (usually for products, but occasionally also for materials) and very frequently in specifications. Advances in science and technology have resulted (and continue to do so) in the development of a better understanding of critical fire properties and of improved methods to assess them. Thus, traditional fire testing procedures are often no longer considered to be the appropriate ones for some materials or products by fire scientists, but are still used, due to commercial interests and the time needed to change requirements.

Ignitability

Unless something ignites, there is no fire. Therefore, low ignitability is the first line of defense in a fire. In fact, however, all organic polymers do ignite, but the higher the temperature a material has to reach before it ignites the safer it is. Ignition temperatures were traditionally assessed with the Setchkin test method for ignition of plastics (ASTM D 1929). Its results are no longer considered very valuable because the exposure is not representative of real fires. However, test results are frequently required and quoted in data sheets, and the test is referenced in building codes for determining suitability of plastics in construction. Ignitability is better assessed via time to ignition or minimum heat input for ignition, on testing for other properties, including using test methods such as heat release calorimeters (OSU, ASTM E 906, or cone, ASTM E 1354). Fire performance improves if either of these is larger. Minimum ignition fluxes for ignition within a certain time period are obtained from the same more modern standard tests for heat release.

Ease of extinction

Once ignited, the easier a material is to extinguish, the lower its associated fire hazard is. LOI, the minimum oxygen concentration required to support candle-like downward flaming combustion, is a measure of the material ease of extinction. It has excellent precision and generates numerical data covering a very broad range of responses. The test (ASTM D 2863) is not an appropriate predictor of real scale fire performance, mainly due to the low heat input and the artificiality of the high oxygen environments used (7), but is widely required in specifications and quoted in data sheets. It is suitable for quality control tool and as a research semi-qualitative indicator of additive effectiveness, if the incident energy is low.

Flame spread

The tendency of a material to spread a flame away from the fire source is critical to fire hazard. Flame spread tests may be applied to polymers or to products in diverse applications or to assemblies. The most widely used tests for specifications and building code requirements are UL 94 and the Steiner tunnel (ASTM E 84, or options for individual products, including NFPA 262 for plenum cables). Other flame spread tests are the radiant panel (ASTM E 162), the lateral ignition and flame spread test (LIFT, ASTM E 1321) and the vertical test for fabrics (NFPA 701). UL 94, for plastics, has vertical and horizontal versions, with varying degrees of severity (vertical tests are more severe than horizontal ones), and is almost universal for plastic materials. The Steiner tunnel test was developed for building materials: wood or gypsum board. It was adopted by every US building code and is the best-known fire test. The specimen is mounted face downward, as the roof of a tunnel and a ca. 90 kW fire source ignites the sample. Unfortunately, the test is used inappropriately for samples that cannot be retained in place above the tunnel floor, or which melt and continue burning on the tunnel floor (typical behavior of thermoplastics), and for thin materials, in spite of giving misleading results for them *(8)*. The normal output is the flame spread index (FSI), a parameter relative to red oak, and the smoke developed index (SDI), the corresponding parameter for smoke obscuration. The radiant panel test has not been shown to be an adequate predictor of real scale fire performance, but results from this test are frequently required in regulations and specifications and quoted in data sheets. The LIFT apparatus is an improvement on the radiant panel test, and determines the critical flux for flame spread, the surface temperature needed for flame spread and the thermal inertia or thermal heating property (product of the thermal conductivity, the density and the specific heat) of the material under test. These properties are mainly used for fire hazard assessment and input into fire models. A flame spread parameter, Φ, is also determined, and this can be used as a direct way of comparing the responses of the specimens. This test method appears well suited for materials (or composites) which are non melting and which can be ignited without raising the incident flux to potentially dangerous limits. It has been used successfully for predictions of full-scale flame spread performance.

Heat release

The key question in a fire is: "How big is the fire?" The fire property that answers the question is the heat release rate. A burning object spreads fire to nearby products only if it gives off enough heat to ignite them. Moreover, heat release must be fast enough not to be lost on travel through the cold air surrounding items not on fire *(9)*. Thus, heat release rate dominates fire hazard, and is much more important than either ignitability, smoke toxicity or flame spread in controlling the time to escape available for fire victims. Three small-scale tests can measure heat release rate: the Ohio State University (OSU), the cone and the Factory Mutual (ASTM E 2058) RHR calorimeters, with the cone the newest and most widely used. It can also assess other fire properties, including ignitability, mass loss and smoke release. Moreover, results from this instrument correlate with those from full-scale fires. For best understanding of the fire performance of materials, it is important to test under a variety of conditions. Therefore test results are often carried out at several incident heat fluxes. Full scale fire test methods have been developed for products, relying on heat release rate measures, including ones for upholstered furniture, mattresses, stacking chairs, electrical cables, plastic display stands, or wall linings. Room-corner tests (NFPA 265, NFPA 286) are now used in codes as alternatives to the Steiner tunnel, as they give more useful results.

Smoke obscuration

Decrease in visibility is a serious concern in a fire, because it (obscuration) results in less light available causing hindrance both to escape from the fire and to rescue by safety personnel. The main way in which a fire causes decrease in visibility is by smoke emission. The decrease in visibility is the combined result of two factors: how much material is burnt in the fire (and less is burnt if the material has better fire performance) and how much smoke is released per unit mass burnt. Empirical parameters have been proposed to compensate for incomplete sample consumption, including the smoke factor, determined in small-scale heat release rate calorimeters. It combines smoke obscuration and heat release rate. The most common small-scale test method for assessing smoke from burning products is the traditional NBS smoke chamber in the vertical mode (ASTM E 662). The test has been shown to have serious deficiencies, mainly misrepresentation of smoke obscuration in real fires and inability to test melting materials, such as thermoplastics *(10)*. Materials that melt or drip generate molten fractions which will have escaped the effect of the radiant heat source, and some of the material does not burn during the test (and does not give off smoke). In real fires molten material will burn and generate smoke. If melting products are exposed horizontally, the entire sample will be consumed. A recent development is the ISO smoke chamber (ASTM E 1995) that allows testing in the horizontal orientation, still in a static environment.

The majority of materials with low flame spread (or low heat release) also have low smoke. However, in 5 recent series of room-corner tests (with the tested material lining the walls or walls and ceiling) heat and smoke were measured and a fraction of the materials tested, ca. 10%, had adequate heat release (or fire growth)

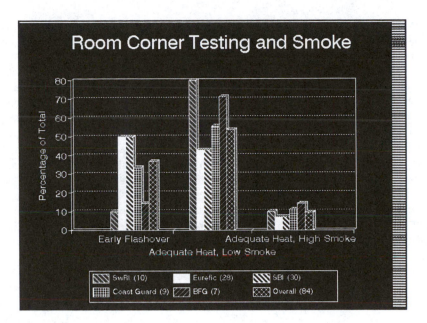

Figure 1: Fraction of products tested in room-corner tests that result in: (a) early flashover, (b) adequate heat and low smoke and (c) adequate heat and high smoke. References: SwRI (USA): (11); EUREFIC (Scandinavia): (12); SBI (European Commission): (13); Coast Guard (Shipboard Use): (14); BFG (USA): (15).

characteristics, but have very high smoke release. In each of the series of tests undertaken, 1 or 2 materials would cause a problem if used in buildings: overall 8 out of the 84 materials tested were found to be severe outliers and have high smoke. The results of all that work are shown in graphical fashion in Figure 1. Work using human volunteers has allowed scientists to conclude that a visibility of 4 m is reasonable for people familiar with their environment *(16)*, so that US codes now include smoke pass/fail criteria in room-corner tests based on that concept.

Smoke toxicity

The majority of fire fatalities are a result of inhalation of smoke and combustion products, rather than the simple effect of burns. Various organizations are trying to develop test methods and guidance documents on smoke toxicity, but emotional responses arise from discussions on interpretation of results or requirements for the use of animals as test surrogates. Most fire fatalities occur in fires that become very large: US statistics indicate that such fires account for over six times more fatalities than all other fires. CO yields in full-scale flashover fire atmospheres (those fires most likely to produce fatalities), are ca. 0.2 g/g, a value translating to a toxicity of 25 mg of smoke per liter and are controlled by geometric variables and oxygen availability, but are virtually unaffected by fuel chemical composition. Small-scale fire tests underpredict CO yields and cannot be used to predict toxic fire hazard in ventilation controlled flashover fires, unless CO yields are corrected. Such tests do not underpredict yields of other toxicants, e.g. HCl or HCN. Toxic potency can be obtained by using small-scale tests. Data from the NIST radiant test are well validated with regard to full-scale fire toxicity, but such validation cannot be done to a better approximation than a factor of 3. Consequently, any toxic potency (LC_{50}) higher than 8 mg/L (toxicity lower than 8 mg/L) will be subsumed within fire atmosphere toxicity and is of no consequence. Fire retardants decrease toxicity in fires, in spite of common misunderstandings that indicate the opposite; the effect is due to a decrease in the amount of burning *(17)*.

Fire Modeling and Fire Hazard Assessment

Overall fire safety is generally achieved by deciding if materials meet certain pre-set objectives. It is generally necessary to combine various properties and calculate results based on fire models, either developed from first principles or based on empirical correlation. Correlation results may generate "safe errors", and be preliminarily acceptable, if predictions suggest that materials are unsafe, when actual full scale testing indicates them to be safe. On the other hand, unacceptable errors are encountered when the test predicts a material is safe, and the full-scale test shows otherwise. Correlation fire models take fire test results, developed in fire safety engineering units, and predict full-scale fire outcomes. The main areas this is used in are: wall linings, seating furniture, and electric cables. Many fire tests have been developed years ago, and are applied beyond what was originally intended. Moreover fire tests may be not be able to generate data that can be used in fire safety engineering applications (such as fire models) and thus simply represent an end point rather than creating a start for additional studies of fire safety. Inappropriate assessment techniques can be: (a) too lenient on materials, and unsafe materials can be introduced into or (b) too severe on the material, so that novel, and suitable, materials may be unfairly excluded from an application where a safe alternative was developed. Fire performance based safety techniques involve fire hazard or fire risk assessment, frequently, but not necessarily, via fire modeling.

An example of the type of results that can be obtained by applying fire hazard assessment concepts to the selection of a new (alternative) product to an existing one follows. The example is the choice of upholstered seating furniture in the patient room of a health care occupancy. The process and its potential outcomes are:

1. Choose fire scenarios.
2. Develop fire safety objectives and apply them to the scenarios chosen.
3. Use assessment techniques (or test methods) as input into calculation methods.
4. Assess the new item of upholstered seating furniture being considered for use in a certain health care facility, and reach one of the following five conclusions.

    i. The new chair is safer, based on fire performance, than the one in established use. Then, the new product would be desirable, from the point of view of fire safety.

    ii. The fire safety of the new and established chairs are the same. Then, there would be no disadvantage in using the new product, from the point of view of fire safety.

    iii. The new chair is less safe, based on fire performance, than the one in established use. Then, the new chair would be undesirable, from the point of view of fire safety.

    iv. The new chair could only be used with suitable fire safety if the design of the patient room is altered, either by changing the layout or geometry, other furnishings or contents, or by adding protective measures, such as automatic sprinklers or smoke detectors, to get equivalent fire safety.

    v. The new chair offers some safety advantages and some safety disadvantages over the one in established use. This could mean more smoke obscuration with less heat release. Then, a more in depth fire hazard assessment would have to be conducted to balance the advantages and disadvantages.

4. If the patient room does not have a chair, then the same fire hazard assessment is done, but the comparison would be to see whether the new chair introduces unacceptable additional fire hazard.
5. The fire hazard assessment reaches a conclusion regarding desirability of the new chair.

The process described above has now become formalized, by adoption into standards (ASTM E 1546, ASTM E 2061), guides (NFPA 555), and regulations (such as those for passenger trains). It is under consideration for the Life Safety Code, building and fire codes and the fire code for ships (NFPA 301). This approach, if properly conducted, ensures a free flow of ideas and the introduction into commerce of innovative designs, which can then compete, on a level playing field with established solutions to a certain need. Its rejection freezes innovation. When a performance based approach is introduced into a certain environment, it is essential, as an interim solution, to retain a prescriptive approach

304

as an acceptable alternative. This would give the necessary leeway to those familiar with the existing system to continue using it.

Final Thoughts

Fire safety of polymeric materials is a combination of a large number of aspects: thermal decomposition, ignition, flame spread, heat release, and smoke obscuration, among others. Many of the issues can be understood on the basis of chemical breakdown and product reaction mechanisms. Polymers have the added problem of being materials that have still not been fully studied either in lab scale or in full scale and that have fire performance characteristics different from those of traditional materials they are replacing. Thus, innovative approaches must be applied to their study, including fire safety engineering techniques. It is critical to realize that safety, and especially fire safety, is not an absolute, but a relative proposition, and that each person, and society as a whole, sets its levels.

References

1.   Cullis, C.F.; Hirschler, M.M. *The Combustion of Organic Polymers*; Oxford University Press: Oxford, UK, **1981**.
2.   van Krevelen, D.W. *Polymer*, **1975**, *16*, 615-620.
3.   Beyler, C.L.; Hirschler, M.M. in *SFPE Handbook of Fire Protection Engineering (2nd Edn)*, DiNenno, P.J. (Editor-in-chief); NFPA, Quincy, MA, **1995**, pp. 1.99-1.119.
4.   Hirschler, M.M. in *Fire Retardancy of Polymeric Materials*, Grand, A.F.; Wilkie, C.; Eds.; Marcel Dekker, New York, NY, **2000**, pp. 27-79.
5.   Lyon, R.E.; Walters, R.N. in *Proc. 5th Int. Fire and Materials Conf., 23-24 Feb. 1998, San Antonio, TX*; Interscience Commun., London, UK, Grayson, S.J. Ed.; **1998**; pp. 195-203.
6.   Hirschler, M.M. in *Developments in Polymer Stabilisation, Vol. 5*; Scott, G. Ed.; Applied Science Publ., London, UK; **1982**, pp. 107-52.
7.   Weil, E.D.; Hirschler, M.M. Patel, N.G.; Said, M.M.; Shakir, S. *Fire and Materials*, **1992** *16*, 159-67.
8.   Belles, D.W.; Fisher, F.L. Williamson, R.B.; *Fire J.*, **1988**, *82*, pp. 24-30, 74.
9.   Hirschler, M.M. in *Heat Release in Fires*; Elsevier, London, UK; Babrauskas, V.; Grayson, S.J.. Eds.; **1992**, pp. 375-422.
10.   Hirschler, M.M., *Fire and Materials*; **1993**, *17*, 173-83.
11.   Hirschler, M.M.; Janssens, M.L. *Proc. 6th Fire and Materials Conf., Feb. 22-23, 1999, San Antonio, TX*; Interscience Commun., London, UK, Grayson, S.J. Ed.; **1999**, pp. 179-198.

12. Ostman, B.; Tsantaridis, L.; Stensaas, J.; Hovde, P.J. *Tratek (Swedish Inst. Wood Technology Research) Report, I 9208053*, Stockholm, Sweden, June **1992**.
13. Sundstrom, B.; van Hees, P.; Thureson, P. *SP Report 1998:11*; Swedish National Testing and Research Institute, Boras, Sweden, **1998**.
14. Grenier, A.T.; Nash, L.; Janssens, M.L. *Proc. 6th Fire and Materials Conf., Feb. 22-23, 1999, San Antonio, TX*; Interscience Commun., London, UK, Grayson, S.J. Ed.; **1999**, pp. 107-118.
15. Hirschler, M.M. in *Proc. Interflam'93*, Oxford, UK, March 30-April 1, **1993**, pp. 203-212.
16. Jin, T. *J. Fire Flammability*; **1981**, *12*, 130-142.
17. Hirschler, M.M. in *Proc. Flame Retardants '94, British Plastics Federation, Jan. 26-27, 1994*; Interscience Commun., London, UK, **1994**, pp. 225-37.

Bibliography of Some Relevant Standards and Codes

* ASTM D 1929: Standard Test Method for Ignition Properties of Plastics
* ASTM D 2863: Standard Test Method for Measuring the Minimum Oxygen Concentration to Support Candle-Like Combustion of Plastics (Oxygen Index)
* ASTM D 5424: Standard Test Method for Smoke Obscuration Testing of Insulating Materials Contained in Electrical or Optical Fiber Cables When Burning in a Vertical Configuration
* ASTM D5537: Standard Test Method for Heat Release, Flame Spread and Mass Loss Testing of Insulating Materials Contained in Electrical or Optical Fiber Cables When Burning in a Vertical Cable Tray Configuration
* ASTM E 84: Standard Test Method for Surface Burning Characteristics of Building Materials
* ASTM E162: Standard Test Method for Surface Flammability of Materials Using a Radiant Heat Energy Source
* ASTM E 662: Standard Test Method for Specific Optical Density of Smoke Generated by Solid Materials
* ASTM E 906: Standard Test Method for Heat and Visible Smoke Release Rates for Materials and Products
* ASTM E 1321: Standard Test Method for Determining Material Ignition and Flame Spread (LIFT)
* ASTM E 1354: Standard Test Method for Heat and Visible Smoke Release Rates for Materials and Products Using an Oxygen Consumption Calorimeter (cone calorimeter)
* ASTM E 1537: Standard Test Method for Fire Testing of Upholstered Seating Furniture
* ASTM E 1546: Standard Guide for Development of Fire Hazard Assessment Standards

* ASTM E 1590: Standard Test Method for Fire Testing of Mattresses
* ASTM E 1623: Standard Test Method for Determination of Fire and Thermal Parameters of Materials, products, and Systems Using an Intermediate Scale Calorimeter (ICAL)
* ASTM E 1678: Standard Test Method for Measuring Smoke Toxicity for Use in Fire Hazard Analysis
* ASTM E 1822: Standard Test Method for Fire Testing of Real Scale Stacked Chairs
* ASTM E 1995: Standard Test Method for Measurement of Smoke Obscuration Using a Conical Radiant Source in a Single Closed Chamber, with the Test Specimen Oriented Horizontally
* ASTM E 2061: Guide for Fire Hazard Assessment of Rail Transportation Vehicles
* ASTM E 2102: Standard Method for Measurement of Mass Loss and Ignitability for Screening Purposes Using a Conical Radiant Heater

* NFPA 101: Life Safety Code
* NFPA 262: Standard Method of Test for Flame Travel and Smoke of Wires and Cables for Use in Air-Handling Spaces
* NFPA 265: Standard Methods of Fire Tests for Evaluating Room Fire Growth Contribution of Textile Wall Coverings
* NFPA 269: Standard Test Method for Developing Toxic Potency Data for Use in Fire Hazard Modeling
* NFPA 286: Standard Methods of Fire Tests for Evaluating Contribution of Wall and Ceiling Interior Finish to Room Fire Growth
* NFPA 301: Code for Safety to Life from Fire on Merchant Vessels
* NFPA 555: Guide on Methods for Evaluating Potential for Room Flashover
* NFPA 701: Standard Methods of Fire Tests for Flame Propagation of Textiles and Films

Chapter 24

Studies of Chemical Behavior in Different Regions of Polymer Combustion and the Influence of Flame Retardants Thereon

D. Price[1], J. R. Ebdon[2], T. R. Hull[1], G. J. Milnes[1], and B. J. Hunt[2]

[1]Institute for Materials Research, Cockcroft Building, University of Salford, Salford M5 4WT, United Kingdom
[2]Department of Chemistry, University of Sheffield, Dainton Building, Brook Hill, Sheffield S3 7HF, United Kingdom

A description is given of a number of techniques established at Salford to investigate chemical behaviour in the various regions of polymer combustion and, also, the various fire conditions under which polymers burn. Laser pyrolysis is used as a model for the so-called 'dark flame' region immediately behind the flame. Conventional cone calorimetry studies are supplemented by chemical analysis taken at various distances from the burning surface, which provides an insight into chemical behaviour through the flame. A whole range of controlled real fire conditions, e.g. well oxygenated, vitiated can be established in the laboratory-scale Purser furnace experiment. In particular, the analysis of the fumes evolved from the polymer fire is facilitated thus providing an insight into their potential toxicity. The value of this combination of techniques is illustrated by results obtained from some polymer and flame retarded polymer systems. Laser pyrolysis studies have provided an insight into the condensed phase flame retardant mechanisms of some methyl methacrylate based copolymers with phosphorus-containing groups incorporated into their structures. Cone calorimetry data are reported for some of these copolymers. The regional variation of the CO_2/CO ratio and the nature of the organic volatiles in a chipboard flame have been studied via chemical analysis of grab samples taken at appropriate locations. CO evolution has been investigated under different fire conditions in the Purser furnace, showing that the combustion toxicity of the polymers studied can be predicted solely from a knowledge of the ventilation conditions.

Polymer combustion is a complex phenomenon involving condensed phase pyrolysis reactions to evolve flammable species, the 'fuel' in any conflagration. This 'fuel' mixes with entrained air and ignites to produce the flame, which feeds heat back to the polymer surface, thus maintaining the 'fuel' supply and establishing a self-sustaining polymer combustion cycle. Smoke and various combustion products, some of which will be toxic, are also released from the flame. Flame retardants act to break this self-sustaining polymer combustion cycle in a variety of ways. The Salford Fire Chemistry Group has established a number of techniques to investigate chemical behaviour in the various regions of polymer combustion and the influence of flame retardants thereon. Laser pyrolysis is used to model the very narrow interface between the flame and the polymer, the so-called 'dark flame' region, see Figure 1. Cone calorimetry is used for conventional studies, e.g. rates of heat release, mass loss and CO evolution, but, in addition, chemical analysis of samples taken from various distances from the burning surface provides an insight into chemical behaviour through the flame. The Purser furnace is a laboratory scale experiment which enables the study of polymer combustion under a whole range of controlled real fire conditions, e.g. well oxygenated, vitiated. In particular, the analysis of the fumes evolved from the polymer fire is facilitated thus providing an insight into their potential toxicity. Figure 1 is a schematic representation of the polymer combustion process and provides an indication of the relevance of the Salford techniques to it. This chapter provides a brief review of these techniques, illustrating their value by means of results obtained from studies of various polymer and flame retarded polymer systems.

Laser Pyrolysis

In a real fire, polymer pyrolysis behind the flame front occurs under extremely high rates of heat transfer. Such conditions are not easily obtained by most available techniques but are similar to the conditions created in some laser-heating processes. The combination of such a laser-heating process with a suitable analytical technique provides a method for the investigation of polymer pyrolysis under conditions analogous to those existing during polymer combustion. This concept was first used by Lum (1) who utilised a continuous laser operated at radiation intensities similar to those occurring in a room fire (ca. 15 J cm^{-2} s^{-1}). The chemical reactions initiated were monitored using a molecular beam-quadrupole mass spectrometry technique. The more recent Salford laser pyrolysis (LP) experiment (2,3) utilises a pulsed laser to heat a solid polymer surface located in the high vacuum region just below the ion source of a time-of-flight mass spectrometer (TOFMS). This high vacuum condition results in a 'freezing' without further reaction of the volatile pyrolysis products as they leave the crater, i.e. the 'reaction zone', produced at the polymer surface (4). Hence the temporal behaviour of the species detected by the TOFMS should be representative of the primary pyrolysis reactions of the polymer. The relevance of this experiment to polymer combustion is best indicated by reference to Figure 1.

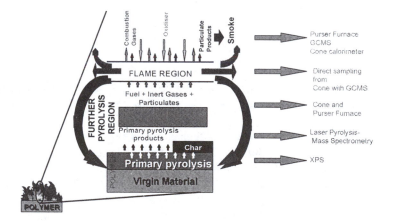

Figure 1. Schematic diagram of the regions of the polymer combustion process and indicating their accessibility to the Salford techniques.

An ongoing collaboration between the universities of Salford and Sheffield is concerned with the synthesis and then quantification and elucidation of the mechanisms of flame retardance in some strategically modified acrylic and styrenic polymers. The copolymers of the phosphorus-containing monomers and methyl methacrylate (MMA) were synthesised via free-radical solution polymerisations as has been described in detail elsewhere (5). The phosphorus-containing monomer typically constituted 10-30% of the monomer mixture. All copolymerisations were taken to about 50% conversion and the MMA-based copolymer then recovered by precipitation in hexane. The copolymers were purified by reprecipitation from solutions in chloroform and dried in a vacuum oven at 50°C for 16 hours. The copolymer studies reported here are those of MMA with pyrocatechol vinylphosphonate (MMA/PCVP), diethyl p-vinylbenzylphosphonate (MMA/DE$_p$VBP) and di(2-phenylethyl) vinylphosphonate (MMA/PEVP). The structures of the monomers are given in Figure 2.

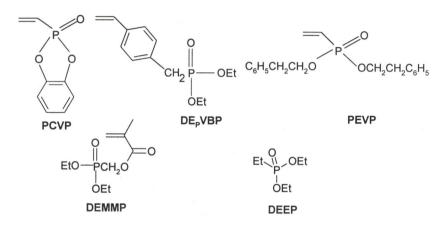

Figure 2. Structures of PCVP, DE$_p$VBP, PEVP and DEMMP comonomers plus that of the additive DEEP.

It is conventional flame retardant wisdom that the more char that is produced on burning the better should be the flame retardance effect. LOI is an empirical measure of flame retardance, i.e. the higher the LOI value the better the flame retardance. However, the data obtained for MMA/PCVP and MMA/DE$_p$VBP are contrary to this. MMA/PCVP has the highest LOI (27.4) but a much lower char yield (5.4%) than MMA/ DE$_p$VBP (12.6%) which has an LOI of 23.6. LP/TOFMS studies were undertaken in the search for an explanation for this observation. The following conclusions were made from inspection of the previously published average mass spectra obtained over the period 600-700 μs after laser pyrolysis of the PMMA, MMA/PCVP and MMA/DE$_p$VBP systems (5).

- Copolymerisation of MMA with PCVP does not significantly change the pyrolysis mechanism compared to that of PMMA. However, there is a reduction in the amount of MMA monomer evolved from the copolymer. This, coupled with 5.4% char formation explains the increased LOI from 17.3(PMMA) to 27.4 (MMA/PCVP). PMMA does not produce char when burnt.

The MMA/DE_pVBP copolymer underwent more extensive decomposition than did MMA/PCVP following laser pyrolysis. In particular, significant amounts of highly flammable methane and ethene were detected. Such increased amounts would also occur if the copolymer were to be exposed to high temperature conditions when burnt. Hence the lower LOI value for the MMA/DE_pVBP copolymer despite it giving a greater char yield than did MMA/PCVP

Another copolymer studied was MMA/PEVP. Again comparison of the average mass spectrum obtained over the period 600-700 μs after the laser pulse with that for PMMA showed that the intensities of the peaks characteristic of evolved MMA, i.e. m/z 100, 69 and 59 are reduced in the MMA/PEVP case (5). Thus copolymerising PMMA with PEVP reduces the propensity of the polymer to evolve MMA when exposed to high temperature as would be the case in a fire, hence the reduction in flammability as indicated by the increase of LOI to 25.5. Also present, at low intensities, in this spectrum are some peaks due to breakdown of the PEVP units, e.g. styrene ions at m/z 104.

Further studies of this copolymer have been carried out using our mass spectrometric thermal analysis technique (MSTA) in which approximately 1.0 mg polymer samples are placed inside a very small tube furnace located in the ion source of the TOFMS and subjected to a 10°C min^{-1} temperature rise. Mass spectra are recorded at regular temperature intervals providing a temperature profile of the evolved products (6). Although it does not produce the rapid heating rates available via laser heating, MSTA has greater sensitivity than LP/TOFMS. Figure 3 shows how the intensities of selected peaks changed as the temperature of the sample was raised. The peaks shown are m/z 100, chosen as a measure of MMA the main 'fuel' evolved in any potential fire, 104 (styrene monomer) and 91 ($C_6H_5CH_2$) characteristic ions of styrene and 122 (the protonated ion from $C_6H_5CH_2CH_2O$) from the di-phenylethyl vinyl phosphonate units.

Consideration of Figure 3 shows that styrene monomer (m/z 104 and 91) and the ($C_6H_5CH_2CH_2O$) unit start to be evolved around 120°C, decreasing to a minimum just above 150°C before increasing again to a maximum at around 190°C. Evolution ceases at about 225°C. Both these species result from cleavage of the ($C_6H_5CH_2CH_2O$) group from the phosphorus atom of the di-phenylethyl vinyl phosphonate units of the copolymer. This would result in the phosphorus remaining within the residual copolymer chain, thus facilitating the subsequent condensed phase action yielding char formation and flame retardance. Evolution of the MMA monomer (m/z 100), on the other hand, only becomes significant just below 200°C reaching a maximum at 250°C. Thus the mass thermal analysis results indicate that the PEVP units decompose at lower temperatures than do the MMA units of the MMA/PEVP copolymer. As a consequence phosphorus atoms remain in the residual copolymer structure facilitating cross-linking and hence char formation as the MMA units begin to decompose to yield MMA monomer as the temperature rises. Thus is the condensed phase flame retardant action of the phosphorus facilitated.

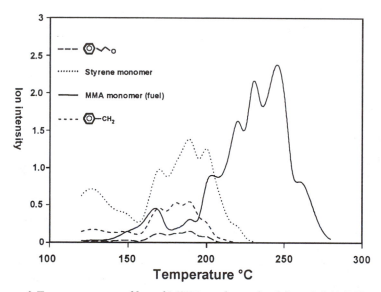

Figure 3 Temperature profiles of MSTA peaks evolved from MMA/PEVP; heating rate 10°C min⁻¹.

Cone Calorimeter

These strategically modified acrylic polymers are also being investigated by cone calorimetry. As an example, we report here the results obtained from plaques produced by copolymerising MMA with diethylmethacryloyloxymethyl phosphonate (DEMMP) and also plaques produced by incorporating the additive diethyl ethylphosphonate (DEEP) into PMMA. These experiments were designed to enable a comparison between an additive and a reactive approach to flame retardance. The structures of DEMMP and DEEP are shown in Figure 2. Previous work had indicated that DEMMP concentrations of 10 mole% or greater give acceptable flame retardance (LOI 22.8) in the MMA/DEMMP copolymer. Thus, for these experiments, the materials compared were MMA/10%DEMMP and MMA+10%DEEP. The glass transition temperatures T_g, as measured via DMTA at 5°C min^{-1}, were as follows: PMMA, 124°C; MMA/10%DEMMP, 117°C; MMA + 10% DEEP, 70°C. These data show that, whereas copolymerisation of MMA with DEMMP has little effect on the T_g value, incorporation of the additive DEEP results in a significant reduction from 124°C to 70°C. Thus the structure of the MMA/10%DEEP is significantly weaker than that of PMMA whereas that of the MMA/10%DEMMP copolymer is essentially the same.

The flame retardance of the materials was assessed by conventional cone calorimetry studies carried out in accordance with ISO 5660 at a radiant heat flux of 35 kW m^{-2}. The data obtained are summarised in Table 1.

Table 1. Cone Calorimeter Data for PMMA, MMA/10%DEMMP and MMA + 10% DEEP

	PMMA	MMA/10% DEMMP	MMA + 10% DEEP
Time to ignition/s	50	60	63
Peak HRR*	641	449	583
Peak CO yield kg/kg	0.0163	0.1748	0.1089
Peak CO$_2$ yield kg/kg	3.903	3.380	2.853
Peak specific extinction area m^2 /kg	155	756	379
Residue/%	1.59	6.59	1.53

Heat release rate/ kW m^{-2}

Whilst the cone calorimeter has been particularly successful in measuring physical parameters of fires, such as rate of heat release (7), it is much less so for chemical studies. The main reason for this is that, in the standard experiment, sampling for chemical analysis is carried out 1 m from the burning sample surface. Also, the fire modelled is that of the fully ventilated condition. As a consequence, samples are greatly diluted by entrained air so that trace species can be difficult to detect. The analyses will be essentially end product analyses since at the flame temperatures involved it is probable that most chemical reactions will be complete 1 m from the flame front. Thus, such analyses would be unlikely to indicate the behaviour of any transient species and thus the reactions involving them. To obtain sensible chemical information for well ventilated fires using the cone it is important that the samples for chemical analysis be taken from much nearer the burning sample surface, i.e. the region where the reactions are actually occurring.

The Salford answer to this problem has been to develop a system for simultaneously taking 'grab samples' from different points in the flame. The technique is illustrated schematically in Figure 4. The samples are taken by opening the solenoid valves, isolating the evacuated vessels, at an appropriate time during the experiment. The timing is normally chosen with reference to the heat release/time profiles. After a predetermined time, necessary for the bulb to be filled, the solenoid valves are closed. The greaseless Young's taps are then closed to seal in the sample. The bulbs are removed and stored at 60°C until the contents are required for analysis. Subsequently, gaseous samples can be removed via the septum using a heated syringe for GC (CO, CO, CO_2 and O_2) or GC/MS (volatile organics) analysis. The bulb is then cooled to room temperature before the condensible compounds can be accessed by introducing 1 ml of solvent, e.g. methanol, via the septum. The surface is then washed around with this solvent before solution samples are removed via the septum for analysis by GC/MS or another chosen method.

For this sampling technique to be successful, it is essential to maintain suitably high temperatures, typically in the region of 200°C, in the sampling lines and the evacuated bulbs during the experiment. The sampling lines should be as short as possible.

The capability of this approach can be illustrated by information obtained from a study of the combustion behaviour of chipboard, a wood-based composite (8). The essential observations were

- CO_2/CO ratio, a measure of the progress of the combustion, changed from 0.5 near the sample surface to 1.0 a further 20 mm away, thereafter increasing rapidly. This variation with distance could explain the lack of correlation reported between the standard cone measurements of this ratio and the ratio values obtained for real fires.
- Unsaturated C_2-C_4 hydrocarbons were the major gaseous organic species present.
- Analysis of the condensibles showed the presence of oxygenated species, e.g. furans, ketones, phenols, etc., mainly near the flaming surface but a little further into the flame. Benzene and derivatives thereof were present throughout the sampled regions of the flame.
- C_9 and C_{11} compounds, particularly naphthalene ring structures which are well known smoke precursors, were significant even at some distance from the burning surface.

Purser Furnace

Yields of toxic products from burning materials vary considerably between the different stages and types of fires. The Purser furnace (9) has been developed as a laboratory scale experiment in which a range of fire types, including those classified by ISO TR 9122, can be produced under controlled conditions. Data on the toxic product yields can be used as inputs to a toxic potency calculation method based on ISO 13344:1996. Thus the Purser furnace is currently being considered as a laboratory scale standard test for predicting the toxicity of materials under different fire conditions (IEC 60695-7-50 and -51). In this respect, it differs from the cone calorimeter which models only one condition, namely fully ventilated under radiant heat. The latter may not be the most toxic burning condition for a particular material.

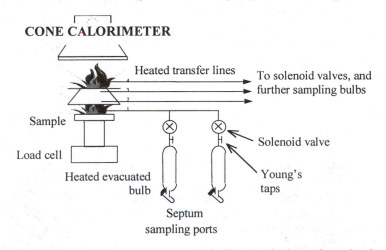

Figure 4 Grab sampling system; sample bulbs attached to solenoid valves via demountable silicone tubing.

As shown schematically in Figure 5, the experiment consists of a tube furnace with a moving sample and controlled temperature and air flow rate. The different fire types are modelled by controlling the temperature and primary air flow in separate experiments. Steady burning conditions can be established over a 10-15 minute period so that a whole range of analytical techniques can be applied to investigate the combustion products, gaseous, liquid and particulate, as well as smoke and residual char. In the Salford version, the hot, undiluted fire effluent is sampled with a paramagnetic oxygen analyser probe (Servomex 570A) whilst the temperature is recorded via a thermocouple. This effluent then passes to a dilution chamber where it is thoroughly mixed with secondary air admitted through a large number of small orifices located at the exit end of the inlet tube. Facilities are available for continuous monitoring of the smoke evolution, CO and CO_2 via infrared absorption, whilst chemical sensors are available to monitor O_2, CO, NO, NO_2 and, if required, HCl and HCN. Fuller details have been published elsewhere (10).

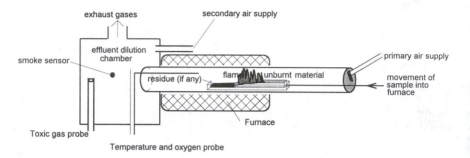

exhaust gases   secondary air supply

effluent dilution chamber   primary air supply

smoke sensor

residue (if any)   flame   unburnt material   movement of sample into furnace

Furnace

Toxic gas probe

Temperature and oxygen probe

Figure 5 Schematic diagram of the Purser furnace.

Purser Furnace Performance

The system has been found to create reliable steady state conditions as illustrated in Figure 6. Ignition occurs about 3 minutes after the start of the experiment. This is the time required for the drive mechanism to push the sample far enough into the furnace for ignition to occur. Steady state burning conditions are established within a minute or so following ignition. A study of a wide range of experimental parameters which affect the performance of a Purser furnace has recently been published (11). It was shown that the furnace, with the temperature set at 750°C, could be used to study a wide variety of fire conditions as detailed in ISO 9122 including unstable, vitiated conditions. The combustion conditions can be set to control the experiment with reasonable accuracy.

One of our current interests is in carbon/oxygen chemistry under different fire conditions. Extensive research, reviewed by Pitts, on the prediction of carbon monoxide evolution from flames of simple hydrocarbons, has shown the importance of the equivalence ratio ϕ, i.e. the actual fuel/air ratio divided by the stoichiometric fuel/air ratio (12). The work has shown that in some circumstances only ϕ is needed to predict the concentration of CO in the products, while in others, the hot or upper layer temperature (the layer just below the ceiling in a room fire), must also be taken into consideration. In both cases there is only a marginal dependence on the fuel. In fuel rich conditions, the CO yield is more or less constant at 0.2 g per g of fuel, for fuels containing only carbon and hydrogen (such as propane, hexane or polyethylene), and about 0.25 g per g of fuel for fuels containing oxygen (such as methanol and propanone). A surprisingly low result for toluene under fuel rich conditions (0.11 g CO per g) was ascribed to its thermal stability leaving more OH· radicals available to oxidise the CO to CO_2.

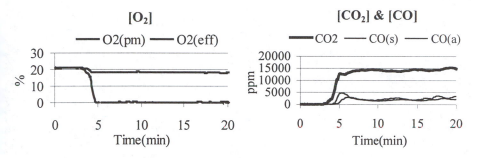

Figure 6   Steady state conditions achieved in the Purser furnace.

In the majority of small-scale fire models, the external heat flux is fixed, and, with the oxygen concentration, controls the rate of burning. It is not possible, for example, to change the oxygen concentration without changing the rate of burning. These models are well suited to replicating the developing fire, in which the CO_2/CO ratio is over 100, but cannot create the more toxic conditions, typical of burning in an atmosphere below 10% O_2. It is generally recognised that small-scale fire tests underpredict CO yields (13). The Purser furnace fixes the fuel feed rate and the oxygen supply, allowing the external heat flux to vary by the depth of penetration into the furnace. It is the only small-scale fire model that provides CO yields and CO_2/CO ratios of similar magnitude to those from fully developed, low ventilation fires, typically 0.2 g/g, showing that the appropriate fire types are being created. Figure 7 presents our Purser furnace data showing the relationship between the equivalence ratio ϕ and the CO yield for a number of polymers containing only C, H and O. Remarkably, the CO yield and hence the combustion toxicity were almost independent of the chemical composition of the fuel for each of polymers studied, except polystyrene.   This implies that the combustion toxicity of any mixture of this group of polymers can be predicted solely from a knowledge of the ventilation conditions.

318

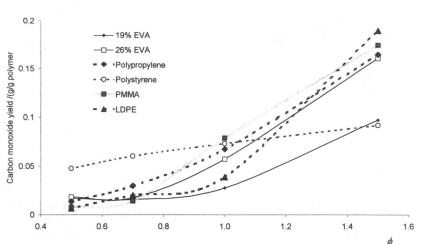

Figure 7 The influence of φ on CO yield for a number of different polymers.

Since the Purser furnace provides data on the oxygen consumption, it is possible to estimate a rate of heat release (RHR) value based on the oxygen depletion principle. It should be emphasised that this is a steady state rate of heat release in contrast to the cone calorimeter measurement, which is a dynamic measurement. Such Purser data would provide an indication of the influence of fire conditions on RHR data. The dependence of rate of heat release from LDPE on ϕ at three different temperatures is shown in Figure 8.

The RHR value varies little with furnace temperature in the range 450-750°C. The value remains roughly the same over the range $0.5 < \phi < 0.9$, i.e. fuel lean conditions, but thereafter falls as the conditions become progressively more fuel rich. Presumably this is because less fuel will be consumed because of the limited oxygen availability.

Summary

The value of the Salford suite of laboratory techniques for studying polymer combustion has been illustrated by information obtained in recent studies of polymer and flame retarded polymer systems.

- Laser pyrolysis provides insight into the condensed phase mechanisms yielding evolved fuel and residual char.
- As well as conventional cone calorimetry measurements, chemical behaviour through the flame can be investigated via analysis of grab samples taken at suitable locations within the burning zone.
- The Purser furnace enables the dependence of the combustion products on fire conditions coupled with the potential to make estimates of their toxic potency.

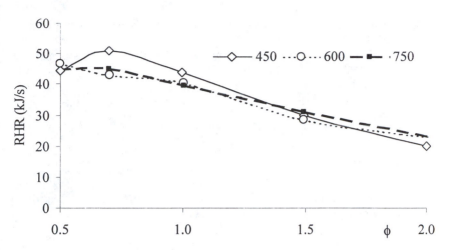

Figure 8 Steady state RHR data for LDPE from Purser furnace at three temperatures (°C); LDPE feed rate 1.0 g min⁻¹.

Acknowledgements

The authors wish to thank EPSRC, ICI Ltd., Albright and Wilson Ltd., and the Fire Research Station for the financial contributions, which have enabled this work to be carried out.   The contributions made by our co-workers, Drs D.A.Purser, W.D.Woolley, F.Gao, L.Lu and P.Joseph, and postgraduates J.A.Carman, K.Pyrah, C.Konkel and L.K.Cunliffe are gratefully acknowledged.

References

1. Lum, R.M., *Thermochimica Acta*, **1977**, *18*, 73-94.
2. Gao, F., Price, D., Milnes, G.J., Eling, B., Lindsay, C.I. and McGrail, P.T., *J. Anal. Appl. Pyrolysis*, **1997**, *40-41*, 217-231.
3. Price, D., Gao, F., Milnes, G.J., Eling, B., Lindsay, C.I. and McGrail, P.T., *Polym. Degrad. & Stability*, **1999**, *64*, 403-410.
4. Price, D., Lincoln, K.A. and Milnes, G.J., *Int. J. Mass Spectrom.Ion Processes*, **1990**, *100*, 77-92.
5. Ebdon, J.R., Price, D., Hunt, B.J., Joseph, P., Gao, F., Milnes, G.J. and Cunliffe, L.K., *Polym. Degrad. & Stability*, **2000**, *69*, 267-277.
6. Price, D., Lukas, C., Milnes, G.J., Whitehead, R. and Schopov, I., *Eur. Polym. J.*, **1983**, *19*, 219.
7. *Heat Release in Fires,* V. Babrauskas and S.J. Grayson (Eds.), Elsevier, England, 1992.

8. Price, D., *Recent Advances in Flame Retardant Polymer Materials,* **1998**, *9*, 38-53.
9. Purser, D.A., Fardell, P.J., Rowley, J., Vollam, S., Bridgeman and Ness, E.M., *Proc. Flame Retardants '94 Conf., London, **Jan. 1994**, Interscience Communications, London, pp. 263-274.*
10. Hull, T.R., Carman, J.M. and Purser, D.A., *Polym. Int.*, **2000**, *49*, 1259-1265.
11. Carman, J.A., Purser, D.A., Hull, T.R. ,Price, D. and Milnes, G.J., *Polym. Int.*, **2000**, *49*, 1256-1258.
12. Pitts, W.M., *Progress in Energy and Combustion Science*, **1995**, *21*, 197-237
13. Hirschler M.M. Proceedings of the 11th BCC Flame Retardancy Conference, Stamford, Connecticut, USA, May, 1999.

Chapter 25

Combustion Toxicity and Chemometrics

Ed Metcalfe[1] and John Tetteh[2]

[1]School of Chemical and Life Sciences, University of Greenwich, Wellington Street,
London SE18 6PF, United Kingdom
[2]DiKnow Ltd., 22 St. James Close, London SE18 7LE, United Kingdom

FTIR spectroscopy was used to monitor gas evolution during combustion. Spectral overlap and other interferences were efficiently modeled by a chemometric methodology called Target Testing. All the important toxic gases can be directly analyzed as a function of time. The results obtained serve as vital input for toxicity modeled such as the Fractional Effective Dose (FED) model, which was used to demonstrate the methodology. Samples studied include wood, PVC and polyurethane foam for which the evolution of key toxic gases has been studied.

Introduction

Toxic fire gases are currently very difficult to measure due to the complexity of the combustion atmosphere and the multiplicity of instrumentation required. To measure the main gases can require various analytical methods including continuous gas monitors such as non-dispersive infrared spectrometry and chemiluminescence, or cumulative methods such as HPLC, GC, ion chromatography, gravimetry, titrimetry. These latter methods do not yield time-dependent data, and only capture integrated values of the concentrations during the combustion process. An important parameter in fire toxicity is the time for the atmosphere to become toxic, which is a function of concentration and the number of gases evolving as a function of exposure time. To capture the true evolution as a function of time, appropriate instrumentation and methods capable of monitoring all the relevant gases are needed.

Fourier transform infrared (FTIR) spectroscopy is a technique capable of monitoring most fire gases, since they are typically small molecules with well-defined infrared spectra, and the technique can be used with a time resolution of a few seconds. However it is not usually possible to use a single wavenumber to monitor a given fire gas, since there are many spectra which are complex and overlap substantially (Figure 1). Using appropriate chemometric methods it should be possible to deal with such complex spectral data sets. Unfortunately traditional multivariate data analysis techniques such as Classical Least Squares, Principal Components Regression and Partial Least Squares are limited in dealing with fire gases due to strict mixture calibration requirements. In this report we used a newly developed multivariate method based on Target Factor Analysis (TFA)(1). This approach eliminates the difficult task of mixture calibration but enables simultaneous monitoring of as many gases as needed. This approach enabled the flexible monitoring of gases for various samples in both real-time and post-run studies. To demonstrate these capabilities combustion profiles of wood, PVC and flame retarded polyurethane foam have been monitored and compared. The results were fed into an FED model to estimate combined toxicity trends as a function of time. The applicability of the methodology to other spectroscopic methods in fire chemistry is also considered.

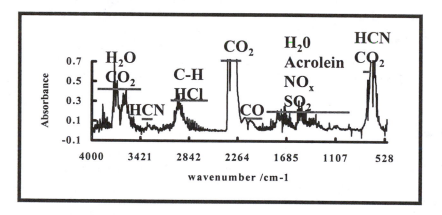

Figure 1. A typical FTIR spectral profile of a gas mixture evolved during combustion of flame retarded polyurethane foam.

Experimental and Data Analysis

Calibration or reference gases are either supplied at known concentrations from standard gas cylinders or can be generated directly from diffusion or permeation tubes (2). Spectra were obtained from standard fire test equipment such as the Cone Calorimeter or the Purser Furnace. Evolved gases were monitored by a Bomem FTIR spectrometer at $4cm^{-1}$ resolution. A full spectrum in the range 4500 to $500cm^{-1}$ was acquired every 7 seconds. All the PTFE sample lines to the heated gas cell were maintained at $150°C$. An average gas flow rate of 4 L/min was also maintained. The gas cell used had a total volume of 0.42 L and a path length of 7.2 meters with KBr windows. Grams32 software by Galactic Industries (4) was used to control the spectrometer to acquire the spectral data. The data was saved as ASCII files and exported to the TFA software developed in the Matlab (5) environment where all the analysis was performed. Details of the chemometric software method have been provided in previous publications (6). A schematic overview of the data analysis process is showed in figure 2.

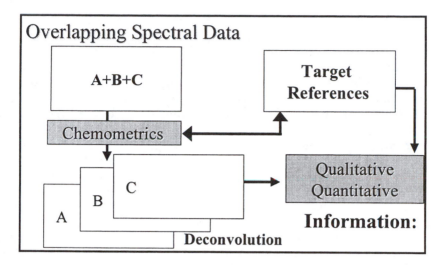

Figure 2. Spectral data block containing gases A, B and C are deconvoluted by chemometrics (Target Factor Analysis) into individual components. Using target references and their known concentrations, the identity (qualitative) and amount (quantitative) information are obtained.

Overview of Target Transformation Factor Analysis (TFA)

TFA is a mathematical technique used to determine whether or not a hypothetical vector, gleaned from chemical principles or heuristic intuition, lies inside the factor space and thus contributes to the phenomenon. A summary of the mathematical theory of factor analysis in chemistry is described in the equations below. More detailed theoretical analysis may be found in reference 1. Essentially the data matrix is first decomposed into abstract factors in row and column space, and the number of significant factors is determined. These significant factors can now be subjected to various form of mathematical scrutiny to directly determine if these factors have real chemical or physical meaning. The number of factor should generally correspond to the number of components absorbing in the spectral window under consideration.

Summary of Data Analysis

A selected wavenumber window of the spectral data was subjected to factor analysis by Singular Value Decomposition (SVD) to obtain abstract factors, also generally called scores and loadings, in both row and column domains of the data matrix. Non-random factors are identified graphically and statistically. This information is used to elucidate the true physical and chemical phenomena contributing to the data and in the process filter out the noise in the data, since the factors arising from noise in the data are separated from those factors that contain significant information.

The key equations are given below. In equations (1)-(4), R^+ is the inverse matrix composed of the significant factors identified after the decomposition of the raw data matrix D into R and C. This matrix is tested with a set of expected factors X_t by a least squares calculation to generate a transformation matrix T. Where a target reference is not known several methods have been proposed in the literature (1) to estimate these unknowns. They include iterative key search and so called needle or uniqueness vector search. The iterative approach used here to estimate X_t where it is unknown is explained in detail elsewhere (1,2). Essentially where a target is not present, a prototype test vector consisting of zeros for all wavenumber points except one or more wavenumber positions where the factor profile show significant absorbance activity is used. The predicted spectrum is resubmitted as a new test vector after putting all negative absorbance values with zero. Typically the spectrum of a likely factor is estimated after five or six iterations. T is then used to operate on R^+ the inverse of the significant factors in the row (spectral) domain to obtain the factors in the spectral domain, X_n. To calculate the profiles in the time domain the least squares fit is performed between the significant factors C in the column (time) domain and the inverse of the transformation matrix T^+ to obtain matrix Y, which contains the evolution (concentration) profiles of the factors.

$$D = RC \tag{1}$$

$$\underline{R}^+ \in X_t = T \tag{2}$$

$$\underline{R} \bullet T = X_n \tag{3}$$

$$T^+ \bullet \underline{C} = Y \tag{4}$$

Modeling Toxicity (e.g. FED)

A typical input configuration for a set of gases to be analyzed is shown in Table I. The table also provides input fields for LC_{50} values required to calculate the fractional effective dose (FED) values. The FED is calculated based on the ISO 13344 protocol (7). The FED represents the toxic effects of various gases as linearly additive:

$$FED = \sum_{j=1}^{n} [Q_{jt}]/LC_{50j}$$

Where Q_{jt} is the quantity (gm^{-3}, or ppmv) of species j at a given time t, and LC_{50j} is the concentration of species j to produce lethality in 50% of test animals within a specified exposure and post-exposure time. The LC_{50} values used in this study based on the ISO 13344 values are listed in Table I. A value of FED=1 means that, at 50% probability level, the lethal level has been reached.

The equation below was used to calculate the FED values presented in this report.

$$FED = [CO]/LC_{50_{(CO)}} + [HCN]/LC_{50_{(HCN)}} + [HCl]/LC_{50_{(HCl)}} + [HBr]/LC_{50_{(HBr)}} + [NO]/LC_{50_{(NO)}} + [NO_2]/LC_{50_{(NO_2)}}$$

Table I. Typical FTIR wavenumber ranges and toxicity data used for automatic gas analysis.

Name	MWt	cm^{-1} [High]	cm^{-1} [Low]	LC_{50}
Acrolein	56.07	1770	1680	
CO	28.01	2180	2110	5700
CO_2	44.01	2410	2390	
HCl	36.46	2850	2600	3800
HCN	27.06	3400	3200	165
NO	30.01	1940	1900	1000
NO_2	46.01	1600	1575	170
SO_2	64.07	1360	1340	
H_2O	18.02	3540	3420	
HBr	80.92	2550	2450	3800
HF	20.01	4200	4000	
Formaldehyde	30.03	1790	1650	
Methane	16.04	3050	2900	
Propane	44.10	3000	2850	
Ammonia	17.03	1210	980	
Methanol	32.04	1080	980	
Toluene	92.14	3150	2850	
Ozone	48.00	1080	1000	
Acetaldehyde	48.05	1800	1680	
Acetic Acid	60.05	1400	1000	

LC_{50} values from ISO 13344 protocol.

Results and Discussion

Figure 3 shows the spectral window for HCN for a sample of flame retarded polyurethane foam. A spectral window of 3200-3400cm^{-1} was chosen for the stretching vibration. Note that the overall absorbances are quite low (<0.08) and there is significant noise in the data, and overlap with other spectra. This complex overlap requires chemometric deconvolution to enable accurate identification and quantification. The first part of the analysis is to identify the number of significant factors. This can be done in three ways: (i) visual inspection of the abstract factors, (ii) the use of the Malinowski indicator function or (iii) by determining the significance level of the successive factors. Full mathematical details of ii and iii are in reference 1. It is recommended that more than one method should be employed to give the best interpretation of the number of factors. In this example there are three factors, one of which is shown by target testing to be HCN. This is demonstrated in Figure 4, which shows the target spectrum from the database, and the resulting predicted spectrum. The correlation coefficient between spectra is 0.932. It is possible that spectra may vary (e.g. broaden) slightly with analysis conditions, and also that the reference spectrum may not correspond to a perfectly pure sample so the fit is very reasonable even though this is quite a challenging deconvolution. The ability to generate estimated spectral profiles through iterative refinement of prototypes permits reasonable estimation of both qualitative and quantitative (concentration trends) content of the data. This is one of the strengths of this

326

chemometric method since it is possible to analyze a set of data without prior information.

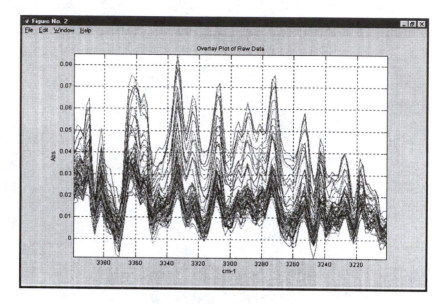

Figure 3. Spectral profile in the 3400-3200cm^{-1} wavenumber region of the flame retarded polyurethane foam

The next stage of the analysis is described in Figure 5. On the left of the figure are shown:
- the spectra of the three factors present,
- the calibration spectra stored for HCN, and
- the quadratic calibration curve for HCN.

Non-linearity is typical for many of the gases analyzed, and we have incorporated quadratic calibration into the model to allow for non-linearity. Since deconvolution is carried out on the basis of the spectral profiles, and concentrations are subsequently calculated from the experimentally derived calibration equations, the predicted concentration values are acceptable, provided that the shape of the spectrum does not change significantly with concentration. This is reasonable if the analysis temperature and pressure are constant. This approach was validated by simulation experiments using known spectra for three overlapping spectra with quadratic absorbance-concentration relationships (unpublished results).

The predicted values for CO and CO_2 in are in good agreement with the NDIR results. For example for PVC in the cone calorimeter, maximum concentrations (where the non-linearity will be greatest) of 1191 and 4884 ppm for CO and CO_2 respectively compare well with NDIR results of 1253 and 5217 ppm at ±5% error margin.

On the right hand side of figure 5 are shown:
- the relative concentration-time profiles for the three factors,

- the extracted spectral data associated with HCN only, and
- the absolute concentration-time profile for HCN.

In effect we have removed the noise and the contributions from the other absorbing components from these data, and the spectra can clearly be seen to correspond to HCN only. There is an initial peak at 50 ppm, followed by a small steady rise at longer times possibly due to HCN release on char degradation.

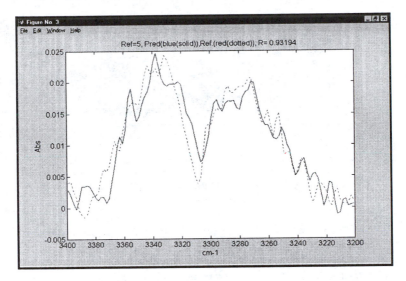

Figure 4. Predicted and reference HCN profile in the selected wavenumber window, showing that HCN is present in the data.

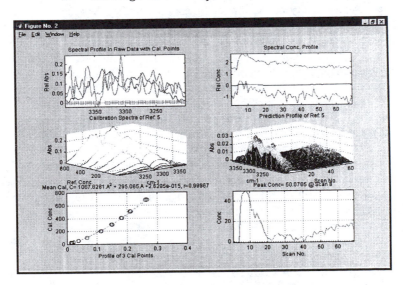

Figure 5. *Summary graphical results showing the various gas evolution profiles in the HCN region. The absolute concentration profile of HCN is shown in the bottom right graph. See text for explanation of the other graphs.*

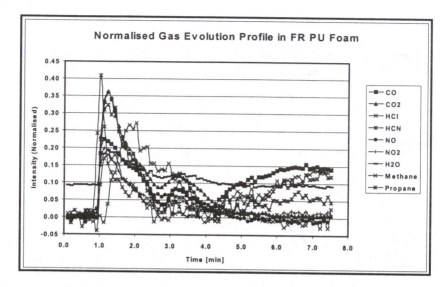

Figure 6. Evolution profile of all target gases in the polyurethane foam. Note that water is present in the before combustion as normal air is used.

Detailed analysis of the individual components demonstrates the different kinetic processes involved for the evolution of the various products. This is shown for several products in Figure 6. The concentrations are scaled to allow comparison by normalization to unit vector. The various components peak at different times, some within a few seconds, others after over a minute. The fire retardant includes a chlorinated additive, and this gives rise to the HCl observed. Both CO and HCN show an additional steady rise at longer times, unlike CO_2, NO and NO_2. Since the latter gases are associated with oxidative, well-ventilated conditions and the former with fuel-rich conditions, it is clear that the later process is associated with degradation reactions within the sample char where oxygen levels are reduced. This additional information can be invaluable in the interpretation of heat and smoke release data from the cone calorimeter and other fire tests, and a useful aid to understanding the nature of the combustion mechanism.

Following the analysis of the spectra from a fire test, the information gained can be tabulated. Table II shows the output from burning wood, PVC and FR polyurethane foam. Identification of target reference is based on a 0.8 minimum correlation coefficient level with the deconvoluted spectrum in the data. The CO and CO_2 data compared well with results obtained simultaneously from non-dispersive infrared analyzers. For PVC the maximum concentrations of CO and CO_2 obtained by NDIR were 1253 and 5217 ppm respectively, by comparison with 1191 and 4884 ppm using the chemometric methodology. The CO/CO_2 ratio is significantly higher for the PVC sample, consistent with the flame retarding effect of the high levels of HCl generated, and consistent with the observation that compounds which do not burn well can generate more toxic atmospheres than those which burn more readily. The HCN, found only from PU foam, reflects the high nitrogen in this polymer.

**Table II. Comparison of gas content of Wood, PVC and PU Foam.
Data from the Cone Calorimeter**

	Max Conc. During Test [ppm]			15min Yield (g)	5min Yield (g)	10min Yield (g)
Gas	Wood	PU Foam	PVC	Wood	PU Foam	PVC
Acrolein						
CO	387	582	1191	1.898	1.280	3.143
CO$_2$	7735	8384	4884	99.709	25.580	19.375
HCl		36	4278		0.110	15.221
HCN		61			0.110	
NO	44	121		0.470	0.240	
NO$_2$	15	12	17	0.221	0.060	0.168
SO$_2$						
H$_2$O	9064	19830	9054	92.335	27.96	38.051
HBr						
HF						
Formaldehyde	98			0.917		
Methane	148	46	212	0.620	0.060	0.383
Propane		16	57		0.03	0.819
Ammonia	26			0.083		

Cumulative yields were taken towards the end of combustion. Sample weights varied. All identified gases were based on correlation coefficients above or equal to 0.8 with the known reference spectral profile.

While for most purposes it may be adequate to analyze the data post-run; it is possible to undertake real time analysis and deconvolution using this methodology. The number of spectra needed to deconvolute the data must be greater than (or equal to) the number of significant factors in the data. In practice we find that for this type of FTIR data, which is inherently noisy because of the low absorbances and the real fluctuations in the fire, between 5 and 10 spectra per factor are needed. This can be clearly seen in real time analysis. Until all the significant factors have been identified, the fit is poor.

Figure 7 shows three target tests for HCl taken on some real data at about 60, 90 and 180 s. After each spectrum is acquired, the data matrix is extended and the data re-analyzed completely.

Initially after only ten scans (61 s.) the fit is poor (r = 0.40), although some structure related to HCl is barely evident in the predicted spectrum. After 15 scans, about 90 s., HCl is clearly indicated as present, but the fit is not strong (r = 0.80). The reason for this is that not all the other factors have been isolated unequivocally. After 30 scans, when all the factors have been located (three in this example), an excellent fit to HCl (r = 0.96) is obtained from target testing. This sequence also illustrates an important rule in chemometrics that one should display results graphically wherever possible. It can be dangerous to have too much confidence in simply numerical data output, and for this reason chemometric software should have some transparency and not be simply a black box calculator. If there is a problem it usually shows up more clearly in graphs, and solutions may more easily suggest themselves (e.g. for a very strong interference, it may be necessary to optimize a spectral window).

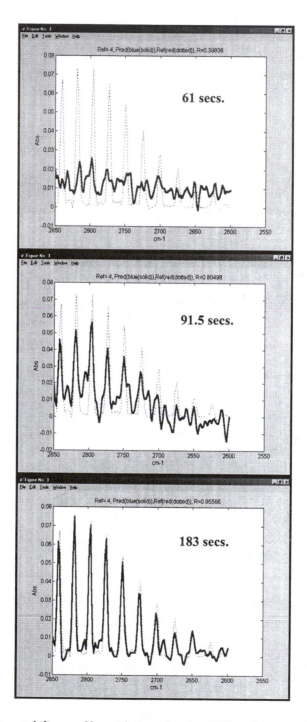

Figure 7. Factor stability profiles with time showing HCl evolution in flame retarded polyurethane foam.

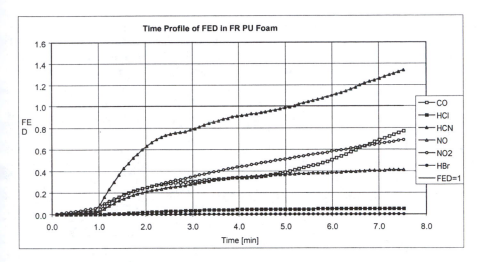

Figure 8. Relative FED profiles for relevant toxic gases identified in the PU foam data, based on scaled-up data from cone calorimetry..

Using, for example, the fractional effective dose model, the concentration-time data, may be converted, by scaling from the test fire to a particular scenario, and incorporating the best available toxicological data. Figure 8 shows the calculated FED profiles for model scenarios for each target gas for which toxicity data are available. The total FED profile may be calculated by summation of the individual curves, allowing the estimation of the time to lethality for the fire atmosphere. This approach is somewhat simplified in that the time scale of the summative FED is different from that of the (30 minute) LC_{50}. Toxicity will be overestimated, since the time scale is less than for the toxicological studies. A more sophisticated model could be developed by assuming a constant exposure time x product concentration for toxicity, i.e. ct = constant. Thus higher concentrations will be needed to generate toxic effects at shorter exposure times. In practice the toxicity behaviour may be even more complex than this, as not all gases obey this relationship, and there may also be interactions between the different gases. The calculation of FED requires assumptions to be made about the fire model in scaling from the test method to a real fire, and the sample size and the enclosure size assumed will affect the calculated FED values. Since the purpose of the present work is simply to demonstrate that the total toxicity of the atmosphere can be generated in principle from FTIR data, the FED values should be considered as relative

We have also used this methodology to study the speciation of phosphorus-based compounds by FTIR (8), during the combustion of tri-methyl phosphate (as a simple model for phosphorus containing flame-retardants such as Reofos) in iso-octane. Complex spectra near 1200 cm^{-1} can be deconvoluted to several components. These may be attributed to di- and mono-methyl phosphate intermediates and to phosphoric acids.

The chemometric methodology described here is not restricted to FTIR, but can be used for other spectroscopic methods. A proviso is that overlapping spectra are linearly additive. This methodology can be used, for example, to study speciation in solutions by UV-visible spectroscopy, and to resolve overlapping mass spectra. In the example given below, we have deconvoluted the laser pyrolysis time-of-flight mass spectrometric data (9) obtained for a PMMA/PEVP copolymer synthesized by Ebdon, Price and co-workers (10). This shows two factors representing successive degradation products on a time scale of < 1ms (Figure 9).

The ability to generate these results is due to the powerful chemometric data analysis employed. This is a significant improvement in the multivariate analysis of fire gases for a very wide range of samples without the need for mixture calibration. By this approach it should be possible to test and evaluate old and new toxicity models and concepts. The characterization of new materials for application to performance and toxicity can now be simplified by incorporating such technology into fire instruments such as the cone calorimeter and others.

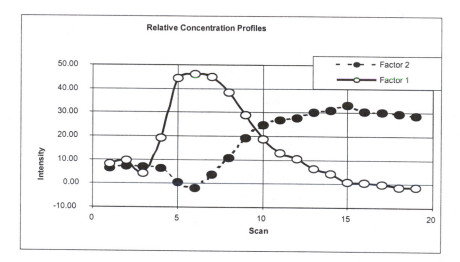

Figure 9. Deconvoluted TOF-MS concentration profiles for laser pyrolysis TOF mass spectrometry of a PMMA/PEVP copolymer (8,9). A two-factor model indicates that two main products are formed.

Conclusions

The analytical flexibility of this chemometric approach is ideal for the practical use of FTIR as a multiple gas analyzer, with high throughput analytical capabilities to profile 20 or more gases. Both qualitative and quantitative information can be obtained rapidly by the application of target factor analytical methods to FTIR spectroscopy for the analysis of fire gases. In cases where no reference information is available spectra and relative concentration profiles are obtained for all absorbing species in a selected wavenumber region. Quadratic calibration enables both linear and non-linear prediction. Recent developments allow many absorbing gases to be monitored, not just by post-run analysis, but also in real time, so that the methodology may have applications in monitoring hazardous environments.

Acknowledgement

Part of the data used in this work was obtained from VTT, Finland through the SAFIR (Smoke Analysis by Fourier Transform Infrared Spectroscopy) project within the European Standards, Measurement and Testing Program under contract number SMT4-CT96-2136.

References

1. Malinowski E. *Factor Analysis in Chemistry,.*, 2nd Edition, Wiley, New York, 1991.
2. Tetteh, J. *Enhanced Target Factor Analysis and Radial Basis Function Neural Network for Analytical Spectroscopy.* PhD Thesis, University of Greenwich, London, UK. 1997.
3. *Generation of test atmospheres of organic vapours by permeation tube methods,* Health and Safety Executive UK, MDHS4, May **1986**.
4. *User Manual Grams*, Galactic Industries 1998.
5. *Mathworks Inc.* USA, Matlab 5.1, 1998.
6. Hakkarainen, T.; Mikkola, E.; Laperre, J.; Gensous, F.; Fardell, P.; Letallec, Y.; Baiocchi, C.; Paul, K.; Simonson, M.; Deleu, C.;Metcalfe, E. *Fire and Materials*, **2000**, *24*, 91-100.
7. *Determination of Lethal Toxicity Potency of Fire Effluents*, ISO 13344, 1996-12-15, First Edition.
8. Smith, C.S.; Metcalfe, E. *Polym.Int.*, **2000**, *49*, 1169-1176.
9. Price, D. Personal communication.
10. Ebdon, J.R.; Price, D.; Hunt, B.J.; Joseph, P.; Gao, F.; Milnes, G.J.; Cunliffe, L.K. *Polym. Degrad. Stab.*, **2000**, *69*, 267-277.

Chapter 26

Thermal Degradation of Poly(vinyl acetate): A Solid-State ^{13}C NMR Study

Caroline M. Dick[1], John J. Liggat[1,*], and Colin E. Snape[2]

[1]Department of Pure and Applied Chemistry, University of Strathclyde, 295, Cathedral Street, Glasgow G1 1XL, Scotland
[2]School of Chemical, Environmental and Mining Engineering, University of Nottingham, University Park, Nottingham NG7 2RD, United Kingdom

The thermal degradation of poly(vinyl acetate) has received little attention in the literature in comparison to polymers such as poly(vinyl chloride). Like poly(vinyl chloride), poly(vinyl acetate) is believed to degrade through elimination of the substituents from the chain backbone to give free acid plus an unsaturated char. Little work has, however, been done on the characterisation of this condensed phase, particularly at large extents of degradation. Thus, in this paper we examine a series of chars by solid state ^{13}C NMR and microanalysis. Samples were prepared by degradation of poly(vinyl acetate) granules in an inert atmosphere over the temperature range 275 - 500°C, corresponding to volatile losses of 7 – 95%. Cross-polarisation, magic angle spinning ^{13}C NMR allows the quantification of changes within the condensed phase both in terms of the loss of acetate functionalities and through development of increasing amounts of sp^2 hybridised carbon. These results, combined with those from dipolar dephasing experiments, indicate that acetic acid loss leads to a cross-linked network containing sp^2 and sp^3 carbons in the approximate ratio 1:1.

334

Introduction

Although polymeric materials now play a central role in modern life, concerns about their contribution to fire deaths remain an obstacle to their still wider use *(1)*. Many fire retardant additives are available but their effectiveness in reducing gross flammability is usually achieved at the expense of increased smoke production *(2)*. Furthermore, these additives, typically based on chlorine, bromine, antimony, zinc or vanadium, are increasingly regarded as environmentally unacceptable *(3)*. Polymers intrinsically capable of producing large quantities of carbonaceous char offer a route to fire retardant systems without these drawbacks. There is a strong correlation between char yield and fire resistance, with char formation occurring at the expense of volatile combustible gases whilst also acting as a barrier against oxygen ingress and volatile product egress *(4)*. Whilst the importance of char is appreciated, the chemistry of its formation has in the past largely been neglected, although the situation is rapidly changing *(5,6)*. In the past, most mechanistic studies of polymer thermal degradation have been solely concerned with volatile products *(7)*, with the residue, often highly cross-linked even in the early stages of degradation, simply regarded as 'intractable'. This principally reflects the limited availability of suitable means of analysing the residue. However, the advent of increasingly sophisticated solid state NMR techniques, many applied originally to structural studies of coal, has provided the opportunity for the direct and detailed characterisation of condensed-phase chemistry *(8-18)*.

Background

The thermal degradation of poly(vinyl acetate) (PVAc) has received little attention in comparison to other polymers e.g. poly(vinyl chloride) (PVC). Like PVC, however, most studies have concentrated on the volatiles produced, in particular the mechanisms of acid loss. The remaining char has received little attention with only FT-IR studies having been reported *(19)*.

Grassie *(20)* conducted the first detailed study on the thermal degradation of PVAc and showed that acetic acid accounted for 95% of the evolved volatiles at degradation temperatures of 213 - 235°C. The remaining 5% of volatiles consisted of carbon dioxide, water and ketene which were suggested to arise from the decomposition of acetic acid. It was established that at such temperatures the polymer loses one molecule of acetic acid per original monomer unit in the chain *(21)*. Subsequent studies by Servotte *(22)* failed to find ketene or carbon dioxide at such temperatures and it was suggested that the copper present in Grassie's experiments was actually acting as a catalyst for acetic acid decomposition. However, studies by Gardner *(23)* using TVA and IR did show their presence.

Grassie *(20)* suggested that the loss of acetic acid and the appearance of colour indicated that the remaining insoluble residue was a high molecular mass polyacetylene. The fact that virtually all the available acetic acid was removed, coupled with the shape of the reaction curves, also led to the suggestion that the acetic acid was not lost randomly but by a chain reaction. It has been suggested that a free radical reaction occurs, which starts at fairly low temperatures and leads to unsaturation of the chain *(19)*.

Although the unsaturation is known to be mainly isolated double bonds and dienes, the presence of small fractions of longer polyene sequences has been reported by Gardner *(23)*. As internal unsaturation increases, crosslinking is promoted. McNeill *(19)* investigated the thermal degradation of PVAc at different temperatures and used FT-IR to analyse the remaining residue. He proposed that crosslinking occurred through macroradical attack on internal substitution, and by Diels-Alder condensation reactions between neighbouring chains, beginning at around 280°C. He also proposed that at temperatures of around 440°C, surviving ester groups formed an α-β-unsaturated ketone through scission of the ester link. FT-IR also suggested the presence of mono-, di-, and tri-substituted aromatic rings at temperatures of 480°C. No information could be extracted, however, about the structure or content of the aliphatic region.

Although these studies have given some insight into the degradation of PVAc, the information on the structure and formation of char is far from complete. In this contribution, solid state ^{13}C NMR analysis has been carried out on a range of chars prepared *ex-situ*. This technique has been shown by our group to provide a further insight into char formation in polymers *(16-18)*.

Experimental

Materials

Poly(vinyl acetate) pellets (Aldrich, M_w *ca.* 83,000) were used for all experiments. Chars were prepared in a purpose-built pyrolysis rig under a nitrogen atmosphere. Plastic tubing at the end of the rig allowed the safe removal of non-condensable volatiles to a nearby fume hood. Each 1.5g sample of PVAc was placed in the reactor and heated at a rate of 10°C per minute to a specific temperature and then held isothermally for 10 minutes before being removed from the furnace for rapid cooling to occur. Temperatures of 275 - 500°C were used, resulting in a range of chars with increasing volatile losses.

Side group cleavage occurs over the range 250 – 375°C, with more extensive fragmentation occurring at the higher temperatures *(19)*. Elemental analysis and solid state NMR were carried out on the virgin PVAc and each of the chars produced.

Solid State ^{13}C NMR

Solid state ^{13}C NMR was carried out at *ca.* 50 MHz on a Bruker DX 200 spectrometer with MAS at 6 kHz. The pulse width used was 3.5µs. Samples of approximately 250 mg were packed inside 7 mm diameter zirconia rotors with Kel-F caps. No background signal was obtained from the rotors or caps. All spectra were processed with a line-broadening factor of 50Hz. For SPE experiments, a recycle delay of 60 seconds was used (the required delay having been determined by a Torchia experiment), and for CP experiments, one of 2 seconds was used. Variable contact time experiments carried out on two of the chars showed the optimum contact time to be 1 ms and this value was used for all subsequent CP experiments. CP experiments were then carried out on the virgin polymer and each of the chars. In order to determine the proportions of protonated and non-protonated sp^2 carbons in each char, dipolar dephasing experiments were carried out using CP with dephasing times of 1 and 45 µs. These generated the total and the non-protonated sp^2 carbon spectra under equivalent conditions. Since protonated carbons have relatively short spin-spin relaxation times ($T_2 < 10µs$) compared to non-protonated carbons ($T_2 > 100µs$), the use of a 45 µs delay for the ^{1}H decoupling channel being switched off represents a reasonable compromise for generating the non-protonated sp^2 carbon spectra. An SPE was also carried out on one of these chars and gave the same sp^2 to aliphatic carbon ratio as the CP experiment indicating that the latter could be relied on as being quantitative.

Elemental analysis

Analysis for carbon and hydrogen was undertaken on a Perkin Elmer Series II Analyser 2400.

Results

Char Preparation

A series of chars (1-9) were generated at 25°C intervals between 275 - 500°C. Table I lists the volatile losses and atomic H/C and O/C ratios of the virgin PVAc and each of the chars. Looking at the C/O ratios, acetic acid loss appears to be complete after 10 minutes at 400°C with < 0.01 C/O, and the resultant char contains C and H in a ratio of 1:1. However, the volatile loss of this char is 81% indicating that some decomposition of the CH network has also occurred as the theoretical volatile loss for acetic acid alone would be only *ca.* 70%.

Table I. Volatile loss, H/C and O/C ratios

Sample	Temperature	Volatile Loss	H/C ratio	O/C ratio
PVAc	-	0 %	1.54	0.51
Char 1	275 °C	7 %	1.53	0.52
Char 2	300 °C	31 %	1.33	0.33
Char 3	325 °C	67 %	1.13	0.13
Char 4	350 °C	72 %	1.06	0.04
Char 5	375 °C	76 %	1.06	0.02
Char 6	400 °C	81 %	1.01	<0.01
Char 7	425 °C	86 %	0.92	<0.01
Char 8	450 °C	92 %	0.75	<0.01
Char 9	500 °C	95 %	0.49	<0.01

Figure 1 shows the results from microanalysis. C and H percentages were determined by analytical methods and the percentage of O present was calculated by difference. These results were used to calculate the H/C and O/C ratios in Table I. One feature that is evident from Figure 1 is the fact that a marked decrease in the %H from the chars prepared at temperatures above 400°C. This is indicative of aromatisation of the chars.

Solid State ^{13}C NMR

CP spectra of the virgin PVAc and some of the chars are shown in Figure 2. Spinning side bands were found to increase with volatile loss and accounted for 10 - 20% of the sp^2 carbon. This was taken into consideration when calculating

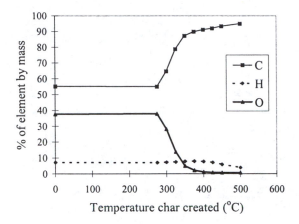

Figure 1. Microanalysis results showing mass % of each element present.

the mole % carbon present in each char. As the samples were spun at ca. 6kHz, the right hand sidebands from the sp^2 carbon were hidden in the aliphatic regions. Their intensities were subtracted from the aliphatic region and added to the relevant sp^2 peak in order to avoid errors in quantification.

Peak assignments for the virgin PVAc are shown in Table II. As degradation proceeds, a decrease in the peaks at 21, 67 and 171 ppm are observed. This is due to the loss of the acetate group. However, the spectrum of char 4 (350°C, 72% volatile loss) still shows a slight peak at 171 ppm indicating that not all acetate groups have been lost. Note that the theoretical value for complete loss of acetic acid is *ca.* 70%; the volatile loss of this char is, however, 2% more than this value indicating that other volatile products arise at such temperatures. As the volatile loss increases, an increase in sp^2 carbon (130 - 140 ppm) is observed. Conversely, we see a decrease in the aliphatic region.

Table III shows the mole % carbon in each sample as calculated from the dipolar dephasing experiments. From these experiments, it was possible to calculate how much of the sp^2 carbon in peak 130-140 ppm was protonated and non-protonated and these values are noted in the table. Acetic acid loss leads to a crosslinked network containing approximately 1:1 sp^2:sp^3 carbon. At higher temperatures, 400°C and above, loss of the aliphatic with ring condensation occurs, leading to an increase in the sp^2:sp^3 ratio. The ratio of protonated to non-protonated sp^2 carbon reaches a limiting value of 2:1 at around 400°C which is indicative of, on average, a network of di-substituted phenyl rings.

340

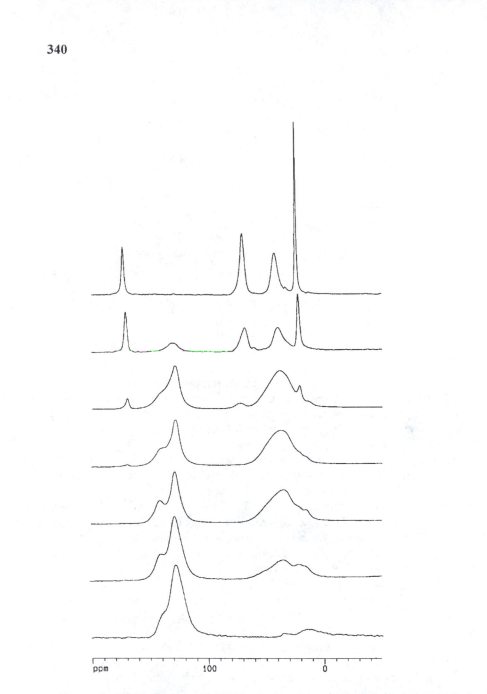

Figure 2. CP-MAS spectra of PVAc (upper spectrum), char 2, char 3, char 4, char 6, char 8 and char 9 (lower spectrum).

Table II. ^{13}C Chemical Shift Assignments for PVAc

δ (ppm)	Assignment
171	C=O
67	C α
40	C β
21	CH$_3$

Table III. Mole percentage of carbon from ^{13}C NMR

Sample	171ppm	130ppm prot	130ppm non-prot	67ppm	40ppm	21ppm
PVAc	23	0	0	26	26	26
Char 2	19	10	1	18	29	23
Char 3	3	33	6	5	47	6
Char 4	1	36	10	<1	54	<1
Char 6	<1	34	18	<1	47	<1
Char 8	<1	44	28	<1	28	<1
Char 9	<1	57	39	<1	4	<1

Discussion

The NMR results are clearly consistent with the established degradation mechanisms of PVAc *(19)*. For example, as clearly shown in Figure 2, as degradation proceeds, a decrease in the peaks at 21, 67 and 171ppm are observed. This is due to the loss of the acetate group. As the deacetylation progresses signals due to sp^2 carbon develop. However, it is interesting to observe that even at 72% volatile loss (350°C) some acetate groups remain; this is consistent with McNeill's results *(19)*. Of more significance, however, is the observation from the dipolar dephasing experiments that the aromatic rings are, on average, disubstituted. This indicates that the aromatic structures may be formed by an intramolecular cyclisation mechanism. This mechanism can be envisaged as occurring via a six-membered transition state consisting of a polyene sequence to give a hexadiene structure that subsequently aromatises via hydrogen transfer.

Conclusions

Solid state ^{13}C NMR has proved to be a powerful tool in the characterisation of crosslinked chars. CP-MAS experiments showed that acetic acid loss leads to a crosslinked network containing approximately 1:1 sp^2:sp^3 carbon. In addition, CP-MAS dipolar dephasing experiments permit the quantification of non-protonated carbon and further aid the characterisation by allowing the identification of crosslinking sites and substitution patterns upon aromatisation.

References

1. Morikawa T.; Yanai E.; Okada T.; Watanabe T.; and Saito Y. *Fire Safety J.* **1993**, *20*, 257-274.
2. Starnes Jr. W.H.; Edelson D. *Macromolecules* **1979**, *12*, 797-802.
3. Lomakin S.M.; Zaikov G.E.; Artiss M.I. *Int. J. Polymeric Mat.* **1996**, *32*, 173-202.
4. Cullis C.F.; Hirschler M.M.; *The Combustion of Organic Polymers;* Clarendon Press: Oxford, 1981.
5. Carty P.; White S, *Polymer* **1994**, *35*, 343-347.
6. *Fire retardancy of polymers – the use of intumescence;* Le Bras M; Camino G.; Bourbigot S.; Delobel R.; Eds.; Royal Society of Chemistry: Cambridge, 1998.
7. Grassie N.; Scott G.; *Polymer Degradation and Stabilisation;* Cambridge University Press: Cambridge 1984.
8. Maciel G.E.; Bartuska V.J.; Miknis F.P. *Fuel* **1979**, *58*, 391-394.
9. Maciel G.E.; Sullivan M.J.; Petrakis L.; Grandy D.W. *Fuel* **1982**, *61*, 411-414.
10. Earl W.L.; Van der Hart D.L. *J. Magn. Res.* **1982**, *48*, 35-54.
11. Maroto-Valer M.M.; Andresen J.M.; Rocha J.D.; Snape C.E. *Fuel* **1996**, *75*, 1721-1726.
12. Maroto-Valer M.M.; Andresen J.M.; Snape C.E. *Energy and Fuels* **1997**, *11*, 236-244.
13. Bourbigot S.; Le Bras M.; Delobel R.; Decressain R.; Amoreux J.P. *J. Chem. Soc., Faraday Trans.* **1996**, *92*, 149-158.
14. Bourbigot S.; Le Bras M.; Delobel R.; Tremillon J.M. *J. Chem. Soc., Faraday Trans.* **1996**, *92*, 3435-3444.
15. Gilman J.W.; Kashiwagi T.; Lomakin S.; Van der Hart D.L.; Nagy V. *Fire Mater.* **1998**, *22*, 61-67.
16. Dick C.M.; Denecker C.; Liggat J.J.; Mohammed M.H.; Snape C.E.; Seeley G.; Lindsay C.; Eling B.; Chaffanjon P. *Polymer International*, **2000**, *49*, 1177-1182.

17. Dick C.M.; Dominguez-Rosado E.; Eling B.; Liggat J.J.; Lindsay C.I.; Martin S. C.; Mohammed M. H.; Seeley G.; Snape C.E. *Polymer*, **2001**, *42*, 913-923.
18. Jiang D.D.; Levchik G.F.; Levchik S. V.; Dick C.; Liggat J.J.; Snape C.E.; Wilkie C.A. *Polymer Degradation and Stability* **2000,** *68*, 75-82.
19. McNeill I. C.; Ahmed S.; Memetea L. *Polymer Degradation and Stability* **1995**, *47*, 423-433.
20. Grassie N. *Trans. Faraday Soc.* **1952,** *48*, 379-387.
21. Grassie N. *Trans. Faraday Soc.* **1953,** *49*, 835-842.
22. Servotte A.; Desreux V. *J. Polymer Sci.:Part C* **1968**, *22*, 367-376.
23. Gardner D.L.; McNeill I.C. *J Thermal Anal.* **1969**, *1*, 389-402.

Chapter 27

Thermal Characterization of Thermoset Matrix Resins

B. K. Kandola[1], A. R. Horrocks[1], P. Myler[1], and D. Blair[2]

[1]Faculty of Technology, Bolton Institute, Deane Road,
Bolton BL3 5AB, United Kingdom
[2]Hexcel Composites, Duxford, Cambridge CB2 4QD, United Kingdom

This work studies the thermal degradation behaviour of a range of commercially available phenolic, epoxy and polyester resins by simultaneous DTA-TGA under a flowing air atmosphere. Thermal analytical transitions of these resins are correlated with their degradation mechanisms cited in literature. Furthermore, TGA studies have also shown that all resins, after decomposition, are completely oxidized and little or no char residue is left above 550°C. Introduction of an inherently flame-retardant cellulosic fibre, Visil (Sateri, Finland), in combination with melamine phosphate - based intumescents (with and without dipentaerythritol) has been made to each resin as a char-promoting system. TGA results of the individual components and their mixtures have indicated that these additives do indeed increase char formation and there is evidence of chemical interaction between different components. A means of increasing the flame retardancy of rigid composites is proposed.

Very recently, interest in intumescent technology (*1*) has been revived by using them as flame retardant additives for polymeric materials. Cellulosic materials are also used in the polymer industry for various applications such as

344

fillers, laminates and panel products. When used in phenolic and other thermosetting resins, they decrease shrinking during molding, give improved impact strength to finished products and are cost effective (*2*). We have recently observed that if certain intumescents are applied to flame-retardant cellulosic fibres, both components interact and develop a unique 'char-bonded' structure which has enhanced flame and heat resistant properties (*3*).

In the present work we have extended the use of intumescents and combinations with cellulosic materials as means of enhancing char formation in thermoset resins. Firstly, thermal stabilities of a range of phenolic, epoxy and polyester resins have been analysed using simultaneous differential thermal analysis (DTA) – thermogravimetric analysis (TGA) up to 1000°C. The various chemical steps of degradation of all the resins have been interpreted using mechanistic studies in the literature. The second part of the research describes the effect of introducing intumescent systems and cellulosic fibres to these resins.

Experimental

Materials

These were selected based on their commercial availability and known chemical characteristics.

Resins :   *Phenolic (Hexcel Composites Ltd) : A1 – A4*
Four phenolic resins differing in phenol/formaldehyde ratios.
Epoxy (Hexcel Composites Ltd) : B1 – B5
B1 is a model formulation with Bisphenol A epoxy resin. B2 is also a model formulation with a multifunctional epoxy resin component. B3 is modification of the B2 formulation with a different epoxy resin. B4, and B5 are modifications of the B1 formulation, again with different epoxy resin components.
Polyester (Scott Bader) : C1 - C4
C1 (Crystic 471 PALV) and C2 (Crystic 2-414PA) are orthophthalic, and C3 (Crystic 491 PA) and C4 (Crystic 199) are isophthalic - based polyester resins.

346

Fiber :   Visil (Sateri Fibres, Finland) - contains polysilicic acid (30% w/w as silica)

Intumescents : Int1 and Int 2 (Rhodia, formerly Albright &Wilson Ltd)
                  Int1 (Antiblaze NW) - contains melamine phosphate and dipentaerythritol in a mole ratio between 1:1 - 2:1.
                  Int2 (Antiblaze NH) - contains melamine phosphate

Equipment

For simultaneous DTA-TGA analysis, an SDT 2960 (TA Instruments) instrument was used under flowing air (100 ml/min) and at a heating rate of 10 K min^{-1}. About 10.0 mg of sample was used in each case. Visil fibres were pulverized using a Wiley mill with a 0.5 mm screen Based on previous experience (*3,4*), the following combinations were studied by simultaneous DTA-TGA to observe the potential interactivity of different components :

- Resin, intumescent and fibre individually
- Resin/intumescent - 1:1 mass ratio
- Cellulosic fibre/intumescent - 1:1 mass ratio
- Resin // fibre/intumescent - 1: 0.5 : 0.5 mass ratio

For the model formulation epoxy resin (sample B1), all the above combinations were studied before and after curing at 135 ^{0}C for 2h. The results showed that there was little difference in their behaviour except for the presence of the curing exothermic peaks for resin at relatively low temperatures. Hence, to simplify experimental procedure, all remaining samples were studied without prior curing.

Results and Discussion

Resin behaviour

DTA and TGA curves for one each for phenolic (A1), epoxy (B1) and polyester (C1) are reported elsewhere (*4*)and DTA and DTG peaks for all resins are given in Table 1. For each resin, the respective transition temperatures are grouped into lower temperature and higher temperature regimes which are resin type dependent.

Table 1. Analysis of DTA responses (peak temp,^{0}C) and DTG maxima (^{0}C) of resins under flowing air.

Resin	DTA peaks		DTG peaks
Phenolic			
A1	140,161 Ex(s,d) ; 526,538,554 Ex(t)		144 527, 541, 556
A2	131,145 Ex(s,d) ; 461,509 Ex(d)		138 440, 460, 515, 543
A3	123,150 Ex(s,d) ; 440 Ex ; 546,563 Ex(d)		101, 126, 140 443, 531, 549, 569
A4	145,164 Ex(s,d) ; 499,538,551,574 Ex(f)		63, 146 501, 540, 580
Epoxy			
B1	160 Ex ; 290 Ex(s) 386 Ex(s) ; 554 Ex		370, 389, 408 567
B2	158 Ex ; 250 Ex(s) 367 Ex ; 553 Ex		366 565
B3	153 Ex ; 248 Ex(s) 370 Ex ; 420 Ex(s) ; 540Ex		366 554
B4	154 Ex ; 357 Ex(s) 432 Ex(s) ; 557 Ex	385	 568
B5	157 Ex ; 290 Ex(s) 357 Ex(s) ; 398Ex ; 549Ex		58, 146 374, 397, 563
Polyester			
C1	193 Exo B 380,395,428 Ex(s,t) ; 524 Ex		360, 378, 394 521
C2	90 Exo B 375,392,410 Ex(t) ; 524 Ex		215, 356 523
C3	90 Exo B ; 385 En 403,424 Ex(d) ; 530 Ex		394 535
C4	96 Exo B ; 331 Ex 391,400 Ex(d) ; 528 Ex		375 537

Key : En = endotherm ; Ex = exotherm ; (s) = small peak ; (d) =double peak ; (t) = triple peak ; (f) = peak with four maxima, Exo B=Exothermic baseline shift

Phenolic resins are manufactured by acid (novolac) or base (resoles) - catalysed reaction of phenol and its derivatives with formaldehyde. The ratio of the phenol to formaldehyde is important. The flammability of phenolics depends on how they are prepared and cured. Flammability increases with an increase in the mole fraction of formaldehyde in the reaction mixture (*5*) and the cured resins are significantly less flammable than the uncured resins.

The first double-peaked DTA exotherms of samples A1-A5 as shown in Table 1, in the range 140-161°C define the curing reaction. According to Christiansen and Gollob (*6*), the first peak (first maximum in the present case) occurs if the resin contains sufficient free formaldehyde and this peak becomes more intense and moves to higher temperature with increasing content. Curing occurs during the second peak (second maximum in the present case), caused by the condensation reactions involving hydroxymethyl groups. The second multi-peaked exotherm above 500°C represents the main decomposition reactions. The first step of thermal degradation is release of water (*7*), which may be because of phenol-phenol condensation (*8*) :

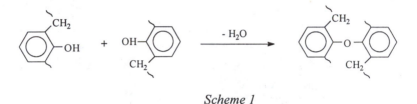

Scheme 1

or alternatively, according to Morterra and Low (from ref 5) via $>-OH + CH_2<$ condensation. The released water then helps in the oxidation of methylene groups to carbonyl linkages (*5*) which then decompose further, releasing CO, CO_2 and other volatile products to yield ultimately char.

In the case of highly cross-linked material, water is not released until above 400°C, and decomposition starts above 500°C (*8*). This is the case for all the samples as shown by DTA and DTG peaks in Table 2. Phenolic resins produce significant amounts of char upon thermal decomposition in an inert or air atmosphere. The amount of char depends upon the structure of phenol, initial cross-links and tendency to cross-link during decomposition (*5*).

Epoxy resins are oligomeric materials that contain one or more epoxy (oxirane) groups per molecule and might contain aliphatic, aromatic or heterocyclic structures in the backbone. Bisphenol-A type resins are most commonly used for composite structures. Epoxy resins are very reactive, hence, both catalytic and reactive curing agents can be used. The general structure of a typical cured epoxy resin can be shown in the Scheme 2, where X can be H and R depends upon the structure of curing agent. Since the catalytic curing agents are not built into the thermoset structure, they do not affect the flammability of the resin. Reactive agents, mostly amines, anhydrides or phenolic resins, on the other hand, strongly affect the cross-linking of these thermosets and hence, their flammabilities (5). Epoxy resins cured with amines and phenol-formaldehyde resins tend to produce more char than acid or anhydride-cured resin.

In Table 1, the first DTA exotherm of sample B1 at 160°C describes the curing reaction. The second DTA and first DTG peaks upto about 410°C represent dehydration and dehydrogenation reactions and above this temperature the peaks are due to the main degradation and char-forming reactions.

During the early stages of the thermal degradation (at lower temperatures) of cured epoxy resins, the reactions are mainly non-chain-scission type, whereas, at higher temperatures chain-scissions occur (9). The most important non-scission reactions occurring in these resins are the competing dehydration and dehydrogenation reactions associated with secondary alcohol groups in the cured resin structures (9):

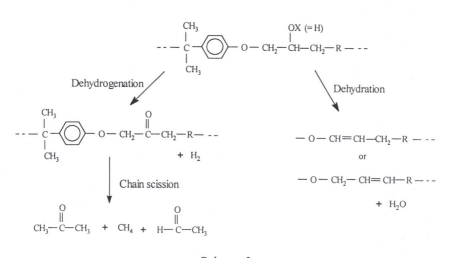

Scheme 2

The main products are methane, carbon dioxide, formaldehyde, and hydrogen. A large amount of methane is also liberated before the start of the above scission reactions following elimination of methyl groups from the bisphenol quarternary carbon as shown in Scheme 3.

Scheme 3

During chain scission reactions the aliphatic segments break down into methane and ethylene (and possibly propylene) or acetone, acetaldehyde, and methane (and probably carbon monoxide and formaldehyde) while from the aromatic segments of the polymer, phenol is liberated. For phthalic anhydride - cured resins, phthalic anhydride is regenerated together with CO and CO_2. Other degradation products are benzene, toluene, o-and p- cresols and higher phenols. In general these are due to further break down or rearrangement of the aromatic segments of the resins.

Aromatic amine - cured resins give large amounts of water in the temperature range 300-350°C (*10*). Rose et al (*11*) have shown by thermal analysis that there are two stages of mass loss. The first step (up to 400°C) mainly involves dehydration of the material and formation of a polyaromatic structure. The second stage corresponds to a thermo-oxidative reaction which leads to complete degradation of the carbonaceous materials.

Polyester resins usually consist of low molecular weight unsaturated polyester chains dissolved in styrene. Curing occurs by the polymerisation of the styrene, which forms cross-links across unsaturated sites in the polyester. Cured resins are highly combustible and burn with heavy sooting.

DTA (*4*) and DTG peaks for samples C1 – C4 in Table 1 upto about 420°C probably represent release of styrene and other volatile products. It is believed that gases evolved during thermal decomposition of polyesters originate from two distinct parts, namely the polystyrene cross-linking and linear polyester components. Most polyesters start to decompose above 250°C, whereas the main step of weight loss occurs between 300 and 400°C (*5*). During thermal decomposition, polystyrene cross-links start to decompose first and styrene is volatilized as shown in Scheme 4.

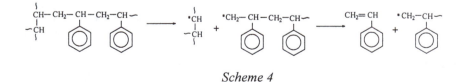

Scheme 4

The linear polyester portion undergoes scission similar to thermoplastic polyesters, undergoing decarbonylation, decarboxylation or splitting off of methylacetylene. Learmonth et al (*12*) have shown that during thermal decomposition volatiles are lost upto 400°C and, above 400°C, it is solid phase oxidation reactions that predominate. DTA and DTG peaks in Table 1, above 500°C are associated with the main decomposition reactions of polyester chains. In support of this conclusion is the similarity of high temperature exotherms for C1 – C4 (524, 524, 530 and 528°C, respectively) which correlate well with respective DTG peak temperatures (see Table 1).

Das and Baijaj (*13*) have studied the degradation mechanism of styrene-polyester copolymers. TGA showed two first order degradation reactions in air. The first one is due to scission of cross-links and weak links with liberation of free linear chains, and the second step is due to random scission of these free linear chains into smaller fragments. Based on our results, these are occurring at the temperatures below 400°C.

Visil fibre, intumescent and their mixtures behaviours

Selected thermal analytical results for Visil, intumescent Antiblaze NH and their mixtures are given in Table 2 to typify mixture behaviors. Thermal degradation mechanisms and thermal analytical results for both the intumescents and their mixtures with Visil have been discussed in detail in our previous communications (*14-16*). Visil on heating releases polysilicic acid which catalyses dehydration reactions of cellulose structure resulting in the formation of CO, CO_2 and ultimately highly cross-linked char. The double peaked exotherm at 325 and 350°C represents these reactions and the other two exotherms represent oxidation of the char. Melamine phosphate on heating undergoes a series of dehydration reactions leading to formation of melamine pyrophosphate (250-300°C), polyphosphate (300-330°C), ultraphosphate (330-410°C) and melam ultraphosphate (410-650°C) followed by decomposition above 650°C (*16,17*). Three endotherms of Int2 (melamine phosphate) at 270, 301, 385°C and exotherm at 685°C represent these dehydration and decomposition reactions, respectively. As discussed before and, as is also evident from Table 2, the results for the mixtures of Visil and intumescents are different from individual components indicating chemical interaction leading to formation of a complex char (*3,14-16*).

Resin - fibre combination behaviour

Resin and Visil mixtures in 1:1 mass ratios showed different DTA and TGA behaviours than expected from component responses. As an example DTA and TGA curves for epoxy resin (B1)/Visil are shown in Figure 1a and b, respectively. The solid lines are the averages calculated from the individual

Table 2. Analysis of DTA responses (peak temp,°C) and DTG maxima (°C) of Visil, Int 2 and their mixtures under flowing air.

	Visil	Int 2	Visil/Int2
DTA peaks	325,350Ex(d) 452 Ex 746 Ex	270 En 301 En 385 En 685 Ex	260 En 321 Ex 374 En 634 Ex 760 Ex
DTG maxima	322 454,494	267,298 383,470 740	258,308,372,438 618 756

Key : En = endotherm ; Ex = exotherm (d) =double peak

component responses. In Figure 1c char yield differences between expected and calculated values from TGA curves for one each for phenolic (A1), epoxy (B1) and polyester (C1) are plotted as functions of temperature and DTG results for mixtures are given in Table 3. These results show that the DTA and DTG peaks for the mixtures are different from those individually for Visil and respective resin and suggest that both the components are interfering with each other's decomposition process. Figure 1c shows that with the phenol resin, the combination is thermally stable as expected upto 450°C, between 450-600°C it becomes less stable than expected, where the char yield difference is about 20% less, and above 600°C there is little difference. For the epoxy resin char is more than expected in the temperature range 350-500°C but less between 500 and 600°C. For polyester, however, residual char is less in the temperature range 250-400°C, slightly more between 400 and 500°C and after that there is not much difference. These results indicate that silicic acid released from Visil may influence char formation to a certain extent and that this char is similarly stable to oxidation at higher temperatures. This suggests that there is little or no interaction between these components with respect to enhanced char production above 500°C of the type observed previously and discussed above for Visil/intumescent combination (*14,15*).

Resin - intumescent and resin - fiber -intumescent interactions

Thermal analytical results for selected combinations are given in Figure 2. and Table 3. DTA behavior in Figure 2 for B1 epoxy resin/Visil/Antiblaze NW combinations indicates that both resin-intumescent and resin-fiber-intumescent

Table 3. DTG maxima (°C) for various stages of decomposition of mixtures of resins with fibre and Int 2 under flowing air.

Resin	Stages	Res/Vis	Res/Int2	Res/Vis/Int2
A1	I	153	136	146
	II	303	259,335,401,557	259,334,399,541
	III	522	809	695
B1	I	318,414,533	166	176
	II		317, 392	326, 380, 514
	III		730	580, 732
C1	I	196	260,320,393	299,352
	II	319	739	776
	III	489		

interactions are occurring. This is corroborated by TGA studies where DTG transitions in Table 3 for resin/Visil/Antiblaze NH show changes for both the two and three – component mixtures with respect to individual component responses (see Tables 1 and 2). This effect is more clearly shown in Figure 3 where char yield differences between expected and calculated values from TGA curves are plotted as functions of temperature. Figs.3(a) and (b) are for combinations with the intumescent Antiblaze NW whereas, Figs.3(c) and 3(d) are for mixtures containing Antiblaze NH, the melamine phosphate component only of Antiblaze NW.

It can be seen from Figs.3(a)-(d) that the two and three component systems are less thermally stable than expected in the temperature range 300 to about 370°C (except for phenolic resin combinations). At higher temperatures, between 400-750°C, they become more stable than expected. Surprisingly, the two component resin-intumescent mixtures in Figs.3(a) and (c) show enhanced char formation above 450°C and the presence of dipentaerythritol in Antiblaze NW does not seem to influence significantly this effect. With regard to resin type, the best results in terms of char enhancement are seen in case of phenolic resin mixtures. However, in the additional presence of Visil (see Figs. 3(b) and (d)) the char formed is still higher than expected and quantitatively similar to two component systems (Figs.3(a) and (c)), although the amount of intumescent present in the former is half that of the latter. Hence, even half the amount of intumescent has produced the same effect in the presence of an equal mass of char-forming fibre. Moreover, the presence of Visil in the three component mixtures results in enhanced char formation above 750°C (see Figs.3(a) and (c)). It should be noted that the effect of silica residue after decomposition of the

354

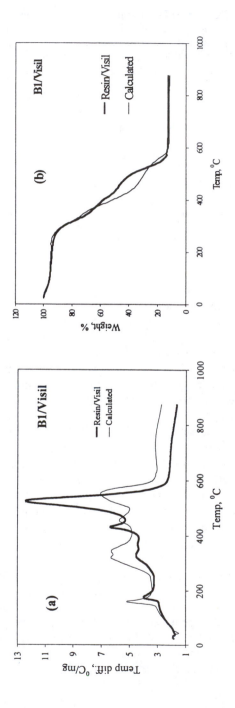

355

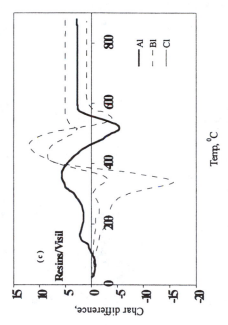

Figure 1. (a) DTA and (b) TGA responses of epoxy resin B1 /Visil combination (_____) their calculated average values (_____) under flowing air. (c) Percentage residual mass difference (actual - averaged) as a function of temperature for phenolic, A1 (_____), epoxy B1(- - - - -) and polyester, C1 (_____) resins.

356

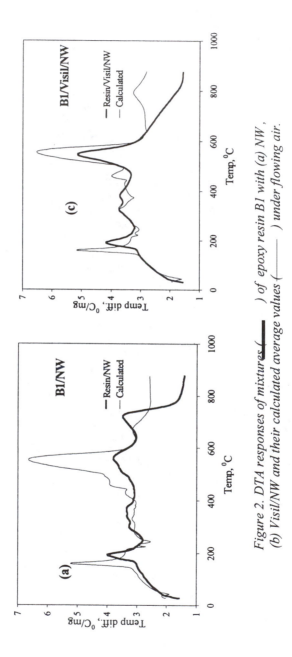

Figure 2. DTA responses of mixtures (――――) of epoxy resin B1 with (a) NW, (b) Visil/NW and their calculated average values (―――) under flowing air.

cellulosic component of Visil is already allowed for and is absent in Figs.3(b) and (d). This indicates that when all three components - resin, Visil and intumescent are present, they interact to form a complex char or "char-bonded" structure which is resistant to oxidation. This effect is also seen in the case of polyester resins, which themselves are not generally efficient char-formers. Thus the presence of intumescent and/or Visil helps in increasing char-forming properties of polyester resins in quite a surprising manner.

DTA and TGA results in Figs. 1 and 3 and Tables 1 and 3, however, also suggest that pure resin degradation chemistries are altered in the resin/intumescent and/or fibre mixtures. Phenolic and epoxy resins degrade initially by dehydration and dehydrogenation reactions (see Schemes 1 and 2) and it is probable that acidic species from the intumescent catalyses dehydration reactions at the expense of chain-scission reactions leading to more char formation at 400°C and above. This synergistic interactive action results in enhanced char formation. While the fiber/resin ratios used here differ form those used in real composites, clearly there are implications for reducing the flammability of composites comprising the addition of a char-forming fibre and an intumescent.

Conclusions

From thermal analytical results there is clear indication of both resin-intumescent and resin-Visil-intumescent interactions during which enhanced residual char formation occurs and even above 750°C for the three component mixtures. It is likely therefore, that melamine phosphate from the intumescent and silicic acid released from Visil, assist in increasing cross-linking of all resins. Subsequently, cellulosic char and resin char-forming reactions interact to form a complex structure which is stable to oxidation. This enhanced char formation indicates that composite laminates produced from these components should have superior flame-retardant properties. Finally it may be proposed that for conventional glass fiber reinforced laminates, pulverized Visil fiber and intumescent may be used as additives or, alternatively, Visil fibers interdispersed with intumescent (either as a fabric or fiber scrim) may be used for partial replacement of glass fiber with possible improvement in flame retardancy. Results of studies of this proposal will be presented elsewhere (18).

Acknowledgements

The authors wish to acknowledge the financial support from the Engineering and Physical Science Research Council and the technical advice and materials supplied by Rhodia Consumer Specialties, Hexcel Composites Ltd., and David Walker Textiles during this work.

358

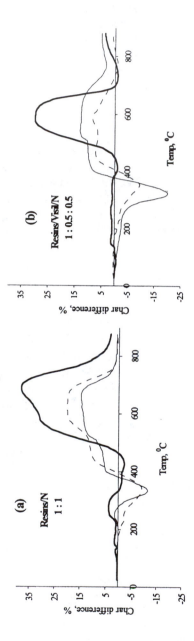

359

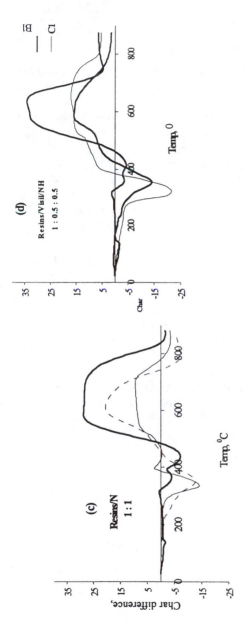
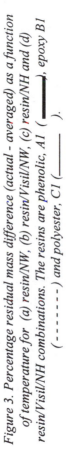

Figure 3. Percentage residual mass difference (actual - averaged) as a function of temperature for (a) resin/NW, (b) resin/Visil/NW, (c) resin/NH and (d) resin/Visil/NH combinations. The resins are phenolic, A1 (———), epoxy B1 (— — — —) and polyester, C1 (————).

360

References

1. Taylor, A.P.; Sale, F.R. *Eur. Polym. Paint Colours. J.* **1992**, *182*, 122-126.
2. Simitzis, J.; Karagiannis, K.; Zoumpoulakis, L. *Eur. Polym. J,* **1996**, *32*, 857-863.
3. Kandola, B.K.; Horrocks, A.R. *Text. Res. J,* **1999**, *69*, 374-381.
4. Kandola, B.K.; Horrocks, A.R.; Myler, P.; Blair, D. in *Proc. Flame Retardants 2000,* Westminster, London, Interscience Communications Ltd., UK, 2000, 217-225.
5. Levchik, S.V. *Fundamentals in International Plastics Flammability Handbook*, 3rd edition, eds., Bourbigot, S.; Le Bras, M.; Troitzsch, J. to be published.
6. Christiansen, A.W.; Gollob, L. *J. Appl. Polym. Sci.*, **1985**, *30*, 2279-2289.
7. Costa, L.; Montelera, L.R.di; Camino, G.; Weil, E.D.; Pearce, E.M. *Polym. Deg. Stab.* **1997**, *56*, 23-35.
8. Jha, V.; Banthia, A.K.; Paul, A. *J. Thermal Anal.*, **1989**, *35*, 1229-1235.
9. Bishop, D.P.; Smith, D.A. *J. Appl. Polym. Sci.*, **1970**, *14*, 205-223.
10. Paterson-Jones, J.C. *J. Appl. Polym. Sci.,***1975**, *19*, 1539-1547.
11. Rose, N.; Bras, M.Le; Bourbigot, S.; Delobel, R. *Polym. Deg. Stab.,* **1994**, *45*, 387-397.
12. Learmonth, G.S.; Nesbit, A. *Br. Polym. J.,* **1972**, *4*, 317-325.
13. Das, A.N.; Baijaj, S.K. *J. Appl. Polym. Sci.*, **1982**, *27*, 211-223.
14. Kandola, B.K.; Horrocks, A.R. *Polym. Deg. Stab.,* **1996**, *54*, 289-303.
15. Kandola, B.K.; Horrocks, A.R.; Horrocks, S. *Thermochim. Acta*, **1997**, *294*, 113-125.
16. Kandola, B.K.; Horrocks, A.R. *Fire Mater.* submitted for publication.
17. Costa, L.; Camino, G.; Luda M.P. *Proc. Am. Chem. Soc.,* **1990**, 211-238.
18. Horrocks, A.R.; Kandola, B.K.; Myler, P.; Blair, D. to be presented at 8th European Conference on Fire Retardant Polymers, Turin, July 2001.

Chapter 28

Flammability Studies of Fire Retardant Coatings on Wood

J. H. Koo[1,*], W. Wootan[1], W. K. Chow[2], H. W. Au Yeung[2], and S. Venumbaka[1]

[1]Center for Flammability Research, Institute for Environmental and Industrial Science, Southwest Texas State University, San Marcos, TX 78666
[2]Department of Building Services Engineering, The Hong Kong Polytechnic University, Hong Kong, China
*Corresponding author: email: jkoo@swt.edu

This paper describes the use of Cone Calorimetry to study fire retardant (FR) coatings based on an industry standard (ASTM E-84) test condition. Several solvent-based and water-based ASTM E-84 Class A FR coatings were tested using the Cone Calorimeter at a radiant heat flux of 50 kW/m^2. The peak heat release rate (PHRR) and total smoke released (TSR) of the Cone Calorimeter were compared to the flame spread index (FSI) and smoked developed index (SDI) of the ASTM E-84 test. A direct correlation between the ASTM E-84 FSI and Cone Calorimeter PHRR was established. This concept enables us to screen FR formulations before conducting the full scale ASTM E-84 test.

Introduction

The function of fire retardant (FR) coatings is to protect surfaces against the extensive heat from fires. When applied to combustible surfaces, FR coatings sharply limit the flame spread and smoke development that would otherwise occur if such combustible surfaces were exposed to fire without a protective coating. To provide a means of determining the surface burning characteristics of combustible materials, the American Society of Testing Materials (ASTM) developed a test standard E-84 (1) to evaluate the fire hazard characteristics of typical construction materials.

It is more cost-effective to use a controlled laboratory device for FR coating product development than to conduct a complete ASTM E-84 test. Numerous researchers have demonstrated this concept in wood products (2) as well as for fiber-reinforced composites (3,4). The Cone Calorimeter enables scientists to conduct ignition tests in a well-controlled and concise manner in the laboratory. This test method was adapted by the US Navy for submarine application (6-8) and was used by numerous researchers to test fire resistant polymeric materials (7-9). This paper describes an attempt to use the Cone Calorimeter (CC) to evaluate FR coatings based on ASTM E-84 test conditions.

Experimental

Description of Materials

We selected several solvent-based and water-based FR coating systems for this study. All of them are U.S. products that possess an ASTM E-84 Class A rating. No compositions were given since these are propriety formulations. An intumescent system containing ammonium polyphosphate (APP) and pentaerythritol (PER) (10) was developed at Southwest Texas State University (SWT).

Material Test Matrix

Nine FR coating systems were tested. Six of them were water-based systems and three of them were solvent-based systems. As a control Douglas Fir plywood (DFP) with no coating was also tested under the same Cone Calorimeter test conditions. Table I summarizes the FR coating ID, manufacturer, coverage rate, and ASTM E-84 results. The thinnest coating is the 20-20 specimen and the thickest coating is the 1500 specimen.

Table I. FR Coating Manufacturer, Coverage Rate, and ASTM E-84 Results

Specimen ID	Manufacturer	Coverage Rate (m^2/L)	ASTM E-84 FSI	ASTM E-84 SDI
Douglas Fir Plywood (DFP)		No coating	55	15
Water-Based				
A-18	No Fire Technologies	4.1	10	55
20-20	Flame Control Coatings	5.6	10	15
1500	Muralo	2.5	15	20
FireCoat 320 (FC320)	Fire Research Laboratory	3.7	10	35
FITEC 104 (104)	Materials Technologies & Sciences	3.6	15	60
Flame Stop II (FSII)	Flame Stop	3.1	25	100
Solvent-Based				
10-10	Flame Control Coatings	4.7	10	20
ClearCoat II (CC2)	Fire Research Laboratory	3.7	20	235
ClearCoat II-A (CCA)	SWT	3.7	NR (no rating)	NR

Preparation of Test Specimens

DFP of dimensions 10- by 10- by 1.8-cm was used as the substrate for all test specimens. A syringe was used to measure and place the amount of coating specified in Table I on the sample. It was evenly spread with a brush that was pre-saturated with the coating. The edges were then fully coated and the coating was allowed to cure at room temperature as specified by the manufacturers. Samples with solvent-based coatings were cured for at least one week before testing.

Results and Discussions

Summary of Cone Calorimeter Data

Parker (11) measured peak incident heat flux levels in the ASTM E-84 Tunnel Test of about 65 kW/m^2. To simulate an average heat flux in the Tunnel Test, a heat flux of 50 kW/m^2 was selected for the Cone Calorimeter. Tables II and III summarize a selective flammability properties of the DFP, the solvent-based, and the water-based FR coating systems, respectively. All data were the mean values of three experiments.

Table II. Cone Calorimeter Data for Solvent-Based FR Coating Systems

Specimen ID	DFP	10-10	CC2	CCA
Sustained ignition time, [s]	14	146	13	9
First peak heat release rate (PHRR), [kWm^{-2}]	245	75	167	124
Time to first PHRR after ignition, [s]	2	64	26	20
Total heat release (THR), [kJ]	1166	960	1146	979
Mass loss percentage	87	78	86	84
Average effective heat of combustion (H$_c$), [MJkg^{-1}]	12	11	12	11
Total smoke released (TSR)	355	121	54	123
CO yield, [kgkg^{-1}]	0.4	0.33	0.5	0.52

Figure 1 shows the heat release rates of the DFP compared to those of the solvent-based FR coating systems. The first peak of the heat release rate of the DFP appeared quite early at 2 s after ignition. The mean PHRR of three experiments was 245 kW/m^2. The second PHHR occurred at 746 s and was 140 kW/m^2. This second PHRR seems to be characteristic of the DFP specimen. This second PHHR was observed in all our FR coated specimens. It shows that the DFP specimen has the highest PHRR, followed by the CC2 specimen, then the CCA specimen; the 10-10 specimen has the best fire performance. Based on Figure 1 and Table II data, CCA performed better than CC2. Since CC2 has achieved an ASTM E-84 Class A rating, the possibility of our experimental formulation CCA to achieve an ASTM E-84 Class A is quite high. It also demonstrated that the amount APP and PER used in the CCA formulation enhanced the performance of the original CC2 formulation.

Table III. Cone Calorimeter Data for Water-Based FR Coating Systems

Specimen ID	DFP	A-18	20-20	1500	FC3200	104	FSII
Sustained ignition time, [s]	14	134	54	176	35	59	16
First peak heat release rate (PHRR), [kWm^{-2}]	245	80	121	70	113	106	186
Time to first PHRR after ignition, [s]	2	87	74	67	13	36	28
Total heat release (THR), [kJ]	1166	1000	1030	973	1041	1058	1148
Mass loss percentage	87	84	81	81	82	81	87
Average effective heat of combustion (H$_c$), [MJkg^{-1}]	12	11	11	10	11	11	12
Total smoke Released (TSR)	355	91	16	70	63	108	26
CO yield, [kgkg^{-1}]	0.4	0.45	0.49	0.35	0.18	0.24	0.39

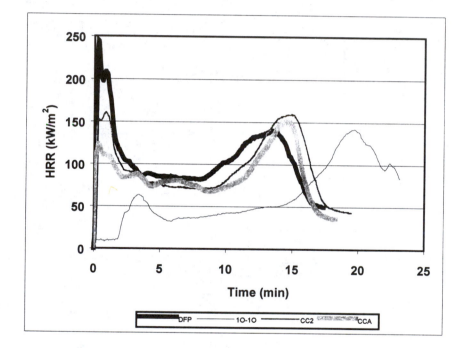

Figure 1. Comparison of HRR for solvent-based FR coatings at 50 kW/m².

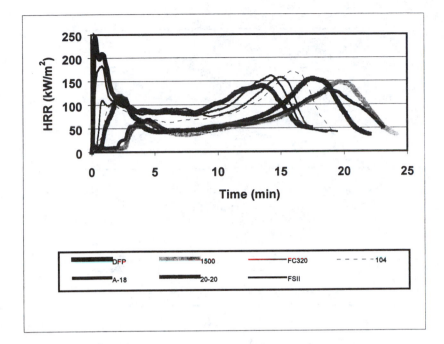

Figure 2. Comparison of HRR for water-based FR coatings at 50 kW/m².

Figure 2 shows the HRR for all the water-based FR coatings compared with the DFP specimen. It shows that the DFP specimen has the highest PHRR followed by the FSII, 20-20, FC320, 104, A-18, and that 1500 has the lowest PHRR characteristic.

Figure 3 shows the HRR for all nine FR coatings. Eight FR coating systems (except CCA) have achieved an ASTM E-84 Class A rating. It is clearly shown in Figure 3, that the HRR of CCA falls within this family of FR coating HRR curves. We have demonstrated this FR coating database is an effective tool for screening experimental coatings before conducting the more expensive full scale of ASTM E-84 test. We can also use this database to obtain an optimal thickness of FR coatings.

Ignitability

Figure 4 shows the ignitability of all the FR coatings. The solvent-based coatings CC2 and CCA both had sustained ignition times lower than the bare DFP. This may be due to residual volatiles that still remained in the solvent-based coating systems. It is quite obvious that even though all the FR coatings

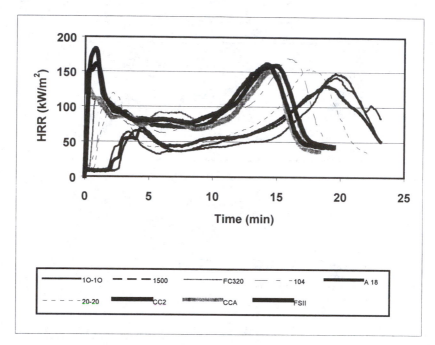

Figure 3. Comparison of HRR for all FR coatings at 50 kW/m².

achieved an ASTM E-84 Class A rating, they varied significantly in Cone Calorimeter fire performance. The sustained ignition time for these FR coatings ranged from 9 s to 176 s. The best three sustained ignition times were obtained by 1500 at 176 s, 10-10 at 146 s, and A-18 at 134 s. The time of ignition of this group of FR coatings is about 2.5 to 3.5 times better than the rest of the FR coatings.

Peak Heat Release Rate

Figure 5 shows that the DFP specimen has the highest PHRR of 245 kW/m². The PHRR of these coatings ranged from 70 to 186 kW/m². The best three PHRR were obtained by 1500 at 70 kW/m², 10-10 at 75 kW/m², and A-18 at 80 kW/m².

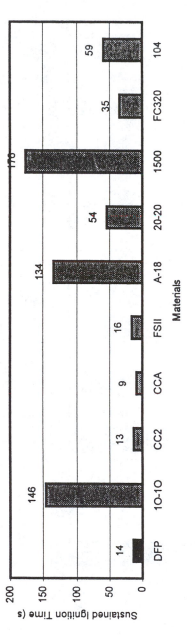

Figure 4. Comparison of sustained ignition time at 50 kW/m².

370

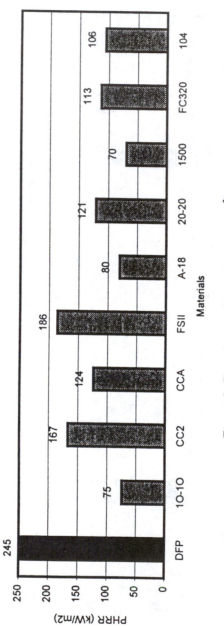

Figure 5. Comparison of PHRR at 50 kW/m².

Smoke and Gas Generation

Tables II and III show that DFP has the highest TSR of 355, the 10-10 solvent-based coating has the next highest TSR of 121, and the 20-20 water-based coating has the lowest TSR of 16. The coated DFP has a much lower TSR (16 to 121) than the uncoated DFP (355). The coated DFP has about the same CO yield as the uncoated DFP. The CO yield is an additional data that we can obtain from the Cone Calorimeter that is not offered by the ASTM E-84 test.

Comparison of ASTM E-84 and Cone Calorimeter Data

Since the ASTM E-84 test duration is only 10 minutes, we re-plotted the HRR for all FR coatings up to 10 minutes. Figure 6 shows the HRR of the Cone Calorimeter data for first 10 minutes to study the FR coatings behavior. It is clearly shown that there are three groups of FR coating performers.

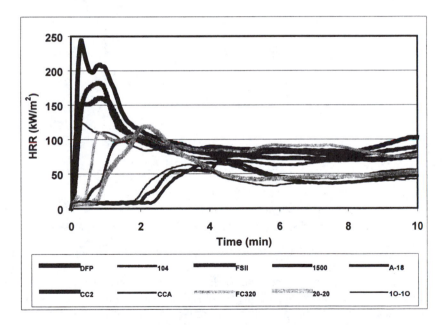

Figure 6. HRRs of FR coatings to simulate ASTM E-84 test duration of 10 minutes.

The first group of FR coatings was able to delay ignition by as much as 2 minutes, then to keep the PHRR at a low level of about 70 - 80 kW/m^2, and to maintain it below 50 kW/m^2 from 6 to 8 minutes. This first group consists of three FR coatings: 1500, A-18, and 10-10 FR coating systems. The second group of FR coatings includes 20-20, 104, and FC320. This group was able to delay ignition only by ¼ to ½ minute, then maintain a PHRR between 106 to 121 kW/m², and keep a steady state HRR of about 70 kW/m². The third group of FR coating was not able to delay ignition, had a PHRR of about 167 to 186 kW/m², and had a steady state HRR of about 75 kW/m². This third group consists of FSII, CC2, and CCA. They are all clear FR coatings.

It is quite obvious that the FR coatings have variable fire performance, even though all had already achieved an ASTM E-84 Class A rating. The Cone Calorimeter was able to subdivide the better performing FR coatings from the mediocre performers.

We tabulated the ASTM E-84 FSI and SDI in Table I; and the Cone Calorimeter first PHRR and total smoke released (TSR) in Tables II and II for all the specimens. Figure 7 shows a comparison of the ASTM E-84 FSI with Cone Calorimeter first PHRR.

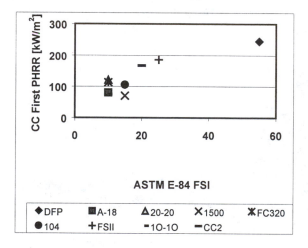

Figure 7. Comparison of ASTM E-84 FSI and Cone Calorimeter PHRR Data

The FSI correlates quite well with the PHRR as shown in Figure 7. This direct correlation between the FSI and PHRR will enable us to screen FR formulations and optimize the coating thickness before conducting an ASTM E-84 large-scale experiment. The ASTM E-84 SDI and Cone Calorimeter TSR did not correlate. This may be due to the difference in time duration of the two experiments. The ASTM E-84 was tested for 10 minutes while the Cone Calorimeter experiments in our study were conducted for about 20 minutes.

Conclusions

We have demonstrated that the Cone Calorimeter is an useful device for screening FR coating systems for product evaluation, research and development. A FR coatings flammability properties database was developed using the Cone Calorimeter. Our experimental CCA coating has a high possibility of achieving an ASTM E-84 Class A rating. Based on the criteria of ignitability and peak heat release rate, the best three FR coatings were 1500, 10-10, and A-18. Since the coverage rate of 1500 is 2.5 m^2/L, 10-10 is 4.7 m^2/L, and A-18 is 4.1 m^2/L, this makes 10-10 and A-18 more efficient than 1500. With the exception of the 10-10 solvent-based coating, the water-based coatings in general perform better than the solvent-based coatings. The Cone Calorimeter enables us to conduct more detailed assessment of FR coatings in a laboratory-controlled and cost-effective manner. The correlation of ASTM E-84 FSI and the Cone Calorimeter PHRR gave us confidence in applying this experimental procedure for our current FR coating development at the Center for Flammability Research.

ACKNOWLEDGEMENT

The authors would like to thank all the fire retardant coating manufacturers for providing free samples. We also thank Ms. Melissa Derrick for preparing the manuscript.

References

1. ASTM E 84. Standard Test Method for Surface Burning Characteristics of Building Materials. American Society for Testing and Materials: Philadelphia, PA.
2. Janssens, M. *Proc First International Fire and Materials Conference.* Interscience Communications: London, 1992, pp 33-42.
3. Stevens, M.; Voruganti, V.; Rose, R. *Proc 10th Annual BCC Conference on Flame Retardancy.* Stamford, CT, 1999.
4. Stevens, M. *Cone Calorimeter as a Screening Test for ASTM E-84 Tunnel Test*; Ashland Chemical Company: Columbus, OH, 1998.
5. ASTM E 1354. Standard Test Method for Heat and Visible Smoke Release Rates for Materials and Products Using an Oxygen Consumption Calorimeter. American Society for Testing and Materials: Philadelphia, PA.
6. MIL-STD-2031 (SH). Fire and Toxicity Test Methods and Qualification Procedure for Composite Material Systems used in Hull,

Machinery, and Structural Applications inside Naval Submarines. Department of Defense: Philadelphia, PA, 1991.

7. Sorathia, U.; Gracik, T.; Ness, J.; Blum, M.; Le, A.; Scholl, B.; Long, G. *Proc International SAMPE Symposium* **2000**, *45(2)*, 1191-1203.

8. Koo, J.; Venumbaka, S.; Cassidy, P.; Fitch, J.; Clemens, P.; Muskopf, B. *Proc International SAMPE Symposium* **1998**, *43(2)*, 1077-1089.

9. Koo, J.; Venumbaka, S.; Cassidy, P.; Fitch, J.; Grand, A.; Bundick, J. *Fire and Materials* **2000**, *24*, 209-218.

10. Lewin, M. In *Fire Retartancy of Polymers: The Use of Intumescence*, Le Bras, M.; Camino, G.; Bourbigot, S.; Delobel, R., Eds.; Royal Society of Chemistry, Cambridge, U.K., 1998; pp 3-32.

11. Parker, W. *An Investigation of the Fire Environment in the ASTM E-84 Tunnel Test;* Technical Note 945; National Bureau of Standards: Gaithersburg, MD, 1977.

INDEXES

Author Index

Subject Index

398